SMASHING REVIEWS FOR

A CHICAGO PUBLIC LIBRARY BEST OF THE BEST BOOK

(Plagues and Pandemics)

AN NSDAR EXCELLENCE IN AMERICAN HISTORY CHILDREN'S BOOK AWARD WINNER

(The American Revolution)

AN NSTA OUTSTANDING SCIENCE TRADE BOOK

(Plagues and Pandemics)

A NOTABLE SOCIAL STUDIES TRADE BOOK FOR YOUNG PEOPLE

(The Mayflower, *Plagues and Pandemics,* and *The Underground Railroad)*

A RISE FEMINIST BOOK PROJECT NOMINEE

(Women's Right to Vote)

★ "Critical, respectful, and engaging."
—*Kirkus Reviews,* starred review on *The* Mayflower

"A multifaceted resource for any school or library."
—*Publishers Weekly* on *The* Mayflower

"Messner proves that she's ready, willing, and entirely able to debunk almost everything you think you know."
—*Booklist* on *The* Mayflower

"Gratifyingly capacious in what it covers."
—*The New York Times Book Review* on *Women's Right to Vote*

"Messner and Meconis provide a timely perspective on an important part of American history."
—*School Library Journal* on *Women's Right to Vote*

"A helpful resource for young researchers."
—*School Library Journal* on *Pearl Harbor*

"An intriguing read."
—*Kirkus Reviews* on *Pearl Harbor*

"Informative and fun, eye-opening and entertaining."
—Chris Barton, award-winning author

"I literally cannot keep History Smashers on my shelves!"
—Lisa Berner, media specialist, Colonie, NY

"My students are LOVING these!"
—Kristen Wolforth Sturtevant, Rochester, NH

"History Smashers are flying off my shelves!"
—Katie Burns Wojtalewicz, Palatine, IL

"Thank you for writing these
'smashers' that create enthusiastic readers!"
—Jenny Volkmann McDonough, Rockville, MD

"History is rarely simple, and Messner does an
outstanding job of making complex topics engaging."
—Kirsten LeClerc, Asheville, NC

"Absolutely love the mixture of graphics and
jaw-dropping facts in these books!"
—Steph McHugh, district book ambassador,
and Mary Hamer, media specialist, Yorkville, IL

THE HISTORY SMASHERS SERIES

The Mayflower

Women's Right to Vote

Pearl Harbor

The Titanic

The American Revolution

Plagues and Pandemics

The Underground Railroad

Christopher Columbus and the Taino People

Salem Witch Trials

Earth Day and the Environment

EARTH DAY AND THE ENVIRONMENT

KATE MESSNER

ILLUSTRATED BY JUSTIN GREENWOOD

RANDOM HOUSE NEW YORK

Visit us on the Web! rhcbooks.com

Educators and librarians, for a variety of teaching tools, visit us at RHTeachersLibrarians.com

Library of Congress Cataloging-in-Publication Data
Names: Messner, Kate, author. Title: History smashers: Earth Day and the environment / Kate Messner; illustrated by Justin Greenwood. Other titles: Earth Day and the environment
Description: First edition. | New York: Random House Books for Young Readers, [2025] | Series: History smashers | Audience: Ages 8–12 | Audience: Grades 4–6 | Summary: "Myths! Lies! Recycling scams? Discover the real story behind the first Earth Day celebration and some of the biggest US climate catastrophes—and their solutions!"—Provided by publisher.
Identifiers: LCCN 2024041048 (print) | LCCN 2024041049 (ebook) | ISBN 978-0-593-70530-8 (trade paperback) | ISBN 978-0-593-70531-5 (lib. bdg.) | ISBN 978-0-593-70532-2 (ebook)
Subjects: LCSH: Earth Day—Juvenile literature. | Environmentalism—Juvenile literature. | Environmental education—Juvenile literature. Classification: LCC GE195.5 .M48 2025 (print) | LCC GE195.5 (ebook) | DDC 394.262—dc23/eng/20240909

The text of this book is set in 13.25-point Napoleone Slab ITC Std.
Interior design by Jade Rector

Printed in the United States of America
10 9 8 7 6 5 4 3 2 1
First Edition

For Meghan Stuart and the readers
of Keeseville Elementary School

CONTENTS

You've probably heard about Earth Day. There's a good chance you've helped celebrate it by planting trees or picking up litter. Maybe you've seen photos of demonstrations and teach-ins from the very first Earth Day, on April 22, 1970, when millions of people came together to teach and learn about how humans have changed the planet. That one-day celebration led to the annual event in April, now celebrated around the world. It also sparked the creation of many environmental groups and inspired people to do more to protect our planet.

But Earth Day wasn't the beginning of environmental awareness—not by a long shot. People didn't just wake up in 1970 and suddenly start caring about nature. Our relationship with the environment goes back a lot farther than that. It's a story about how we use resources, from land to water to energy, and how we dispose of waste. A story of how we protect wildlife and wilderness and what happens when we don't. It's a story of honest mistakes, brazen lies, and solutions we're still trying to figure out.

The real deal about Earth Day and the environment goes beyond simple calls to reduce, reuse, and recycle. It goes all the way back to the first people who lived on the planet and continues with all of us who live here now.

One
THE EARLY ENVIRONMENTALISTS

Humans have lived on Earth for hundreds of thousands of years. For most of that time, they hunted animals and collected edible plants for food. The earliest hunter-gatherers lived in Africa and didn't have much of an impact on the larger environment. Remember, there were no cars or airplanes or even scooters back then. Most early hunter-gatherers were nomadic, traveling throughout the year to follow their food sources, and didn't take more than they could use. But eventually things started to change.

Fire was the first big turning point. Once people harnessed its power, they could cook their food and travel greater distances, as fire provided light, warmth, and protection from predators. They also began using fire to clear trees and brush. Suddenly, humans could change their landscapes on a larger scale.

Another shift occurred when people started moving out of Africa to other continents. Instead of just hunting and gathering, they began to raise their own animals and grow their own plants. More reliable food sources allowed humans to create more permanent communities. It also meant they had a greater impact on the land. Over time, people in many of these ancient civilizations—from Africa to Asia to South America—came to understand that certain ways of farming were easier on the land than others. They were among the world's first environmentalists.

Ancient Maya people, for example, lived in southern Mexico and Central America thousands of years before Europeans came to the Americas. Modern researchers have studied soil samples from the Maya city of Yaxnohcah, on Mexico's Yucatán Peninsula. Those samples, which date from 800 BCE to 1200 CE, show that the Maya were growing a variety of plants, including corn, cotton, avocados, plums, and other fruits.

The Maya used a technique now known as slash-and-burn farming, where people would burn a forest and then grow crops in the rich soil left behind. This technique works for a while but has a negative impact on the environment over time. Trees and other plants help keep dirt in place; once they're burned, that rich soil has no protection from strong winds and rain and can erode, or wash away. Not the best strategy when you have a growing city with more and more people to feed.

So the Maya then began building terraces for farming—tiers of land that would produce more crops and reduce erosion. Preserving the tropical forest around the fields also helped keep water from draining too quickly.

The Maya also built reservoirs to store water from the seasonal rains, saving it to use for drinking and crops when the weather was dry. This practice, which lengthened their growing season, is an example of sustainable land management.

WHAT IS SUSTAINABILITY?

We use the word "sustainable" to describe a practice that will work consistently over the long term. Sustainable agriculture

involves farming practices that don't wipe out the nutrients in the soil. Sustainable energy sources include the sun and the wind, which are always around, as opposed to fossil fuels, such as coal and oil, which took millions of years to form and won't be available anymore once they're used up.

Before Europeans colonized the Americas, people from many Native groups used and managed the land in ways that were largely sustainable.

The Pequots, Narragansett, Wampanoag, and other tribes understood that it was never good to depend on just one kind of food. They planted corn, beans, and squash together, in a way that helped all three plants grow. These Three Sisters, as the crops were called, were the foundation of their diet.

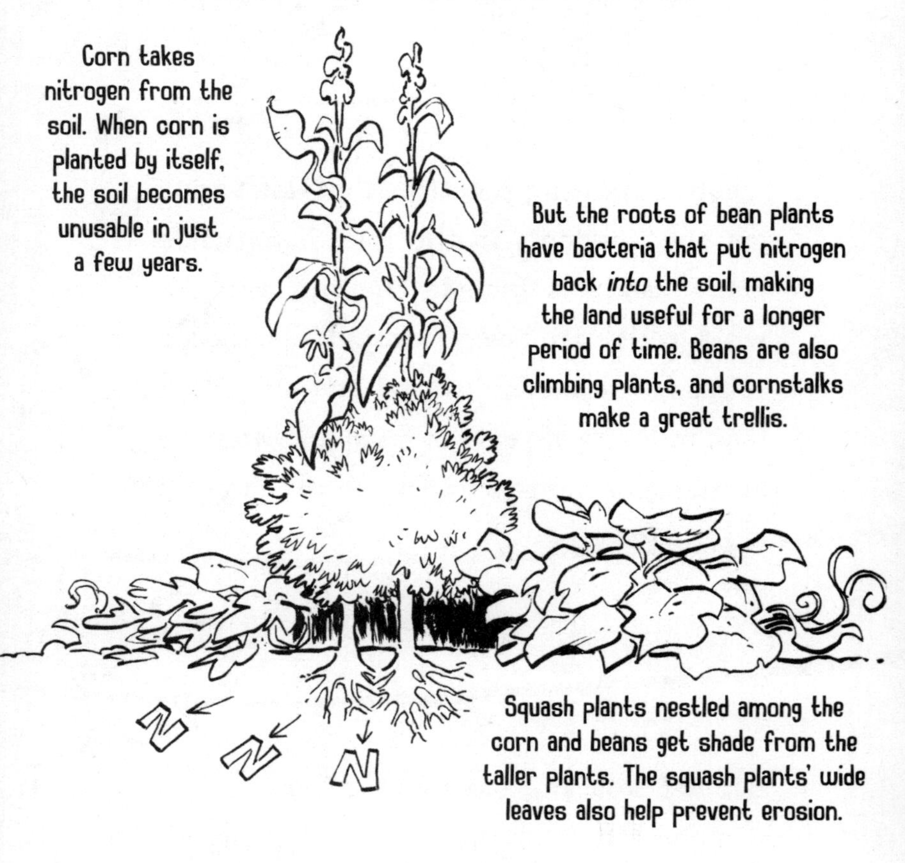

These Native farmers were also hunters and gatherers, depending on the season. They traveled to fishing grounds when bass and salmon were spawning in the rivers. They hunted deer, turkeys, rabbits, and moose for fat and protein and used other parts of the animals for everything from tools to blankets and storage containers. They were careful not to take more than they needed so animal populations would have time to bounce back.

Native people rotated land use, switching up the crops they planted in fields each season to give the soil time to recover. Their land included a mix of meadows, farm fields, and forested areas, all of different ages. This mostly sustainable way of life worked for thousands of years.

When European colonizers began arriving in the fifteenth and sixteenth centuries, they didn't understand Native agricultural practices. In fact, they claimed that the Native people's failure to use all their land was grounds for taking it.

SCHOOLS OF COD OFF THE NEW ENGLAND COAST
WERE SHIPPED OVERSEAS AND SOLD IN ENGLISH MARKETS.

AT FIRST IT SEEMED AS IF THE
RESOURCES WERE ENDLESS.
BUT AS SETTLERS CLEARED MORE
AND MORE LAND FOR FARMING, IT
BECAME OBVIOUS THAT THEY WERE
ALREADY HAVING AN IMPACT.

IN VIRGINIA, TOBACCO
FIELDS THAT HAD REPLACED
NATIVE FARMS SUCKED
NUTRIENTS FROM THE SOIL.

None of these practices were sustainable. And the colonists did eventually notice that when you run around using resources without much thought, those resources don't last. That's when they started making some laws.

In 1668, the Pilgrims, who colonized the area around what is now Plymouth, Massachusetts, drafted one of the first written descriptions of water pollution in the Americas.

Whereas great complaint is made of great abuse by reason of fishermen that are strangers who fishing on some of the fishing ground on our coast, in catches dressing and splitting their fish aboard, [throw] their garbage overboard to the great annoyance of fish which hath any may prove greatly detrimental to the country; it is ordered by the Court that something be directed from this Court to the Court of the Massachusetts to request them to take some effectual course of the restraint of such abuse as much as may be.

In other words: People fishing off our coast keep throwing garbage and fish guts overboard. It's making the water gross and unhealthy for fish. That's really going to mess things up here, so we need to write to the Court of Massachusetts and get it to do something about the problem.

William Penn, who founded Pennsylvania, enacted a forest conservation law when he arrived in the colony in 1682. It required settlers who received land grants to conserve one acre of trees for every five acres they cleared.

Benjamin Franklin may be best known as an inventor, printer, and diplomat, but he was also an early environmentalist in some ways.

A painting of Benjamin Franklin from 1778

Franklin was interested in energy conservation. He developed the Franklin stove, which produced more heat with less fuel. He also joined other citizens in urging lawmakers to stop tanneries, workshops that processed animal hides, from dumping waste into the public water supply. But in colonial America, such concern over natural resources was the exception rather than the rule.

DRESSED TO KILL

European colonizers were excited about the wonderful animals in the Americas. They especially loved beavers. Well, their pelts anyway. Beaver fur has up to a hundred thousand hairs per square inch. (By comparison, you probably have eight hundred to twelve hundred hairs per square inch on your head!)

So beaver pelts are dense, pliable, and waterproof—perfect for the wetlands where beavers live. And also perfect for hats.

Beaver top hats became such a fashion statement in Europe that colonists couldn't keep up with demand. They traded with the Haudenosaunee, Wyandot, Conestoga-Susquehannock, and other Native groups, offering European goods like tools and pots in exchange for beaver pelts. The beaver population plummeted.

Later, European explorers on America's west coast noticed that sea otters also had nice fur. They began hunting otters and selling their pelts to China, a practice that led to the near extinction of these animals.

Today, sea otters are protected to prevent such overhunting, and populations have begun to recover.

Sea otters are found along the west coast of the United States and Canada.

Bison share a similar sad story. Native people of the Great Plains depended on the bison for centuries. But after Europeans brought horses to the Americas, bison could be hunted on horseback instead of just on foot. Once bison hides were introduced to European markets, there was a huge incentive to kill more animals, and the species nearly went extinct.

It's not that Native people were perfect caretakers for the environment before Europeans showed up. People of many tribal nations burned forests to facilitate hunting and farming. Sometimes they stampeded entire herds of bison over cliffs and killed way more animals than they needed. They certainly had an impact on the land and wildlife—but it was on a far lesser scale than what would come after the arrival of Europeans.

And while Native societies were incredibly diverse, they shared a common respect for the environment. They relied on forests, streams, lakes, and soil to survive, so it made sense that they had an appreciation for nature. This gratitude was reflected in their religious belief that the spirit of nature inhabits everything. (This teaching is also part of other faiths, such as Hinduism and African traditional religions.)

European colonizers had different core beliefs. Christopher Columbus, who explored the area now called the West Indies, and the Puritans, who colonized what is today Massachusetts, were Christians. They believed God had given them permission to use the earth's resources however they wanted. It said so right there in the first book of the Bible.

"And God blessed them, and God said to them, Bring forth fruit, and multiply, and fill the earth, and subdue it, and rule over the fish of the sea, and over the fowl of the heaven, and over every beast that moveth upon the earth. And God said, Behold, I have given unto you every herb bearing seed, which is upon all the earth, and every tree, wherein is the fruit of a tree bearing seed; that shall be to you for meat."

—FROM THE GENEVA BIBLE (THE TRANSLATION THE PURITANS READ)

These colonizers believed that people were at the center of all creation. They thought it was not only their right but their *duty* to control the land they found, as well as any non-Christian people who lived there before.

And that belief shaped everything that happened next.

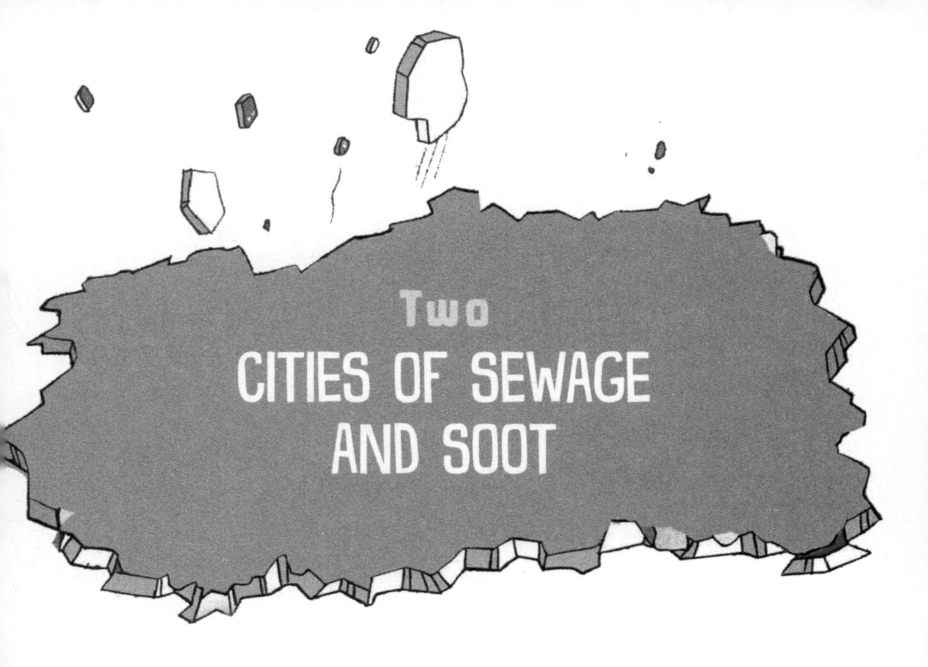

Two
CITIES OF SEWAGE AND SOOT

European colonizers believed in an idea called manifest destiny. They were sure God wanted them to spread their civilization and culture across the land. After the American Revolution, when the United States became an independent nation, they did just that.

Populations skyrocketed and cities grew. Industries popped up in the East. Farms spread across the South and the Midwest, ranches and orchards in the West. All this expansion required land, which colonizers either bought (often at unfair prices) or stole from Native people. Then they forced the Indigenous communities off the land where their ancestors had lived for thousands of years. And European immigration only continued to increase.

From 1851 to 1860, more than 2.5 million immigrants arrived. This growth, combined with advances in transportation and technology, led to America's Industrial Revolution.

Cotton mill workers in North Carolina, 1908

FULL STEAM AHEAD

Across the Atlantic, Britain's Industrial Revolution was already well underway. It got a kick start when James Watt invented a more efficient steam engine in the 1760s.

Watt has been called the Father of the Industrial Revolution. His new design needed less coal to create more power. However, the new steam engine became *so* popular that it resulted in a lot more coal being burned overall. That meant sooty air in cities. And even though people didn't realize it at the time, lots of carbon dioxide was being pumped into the atmosphere, where it would later cause the climate to warm.

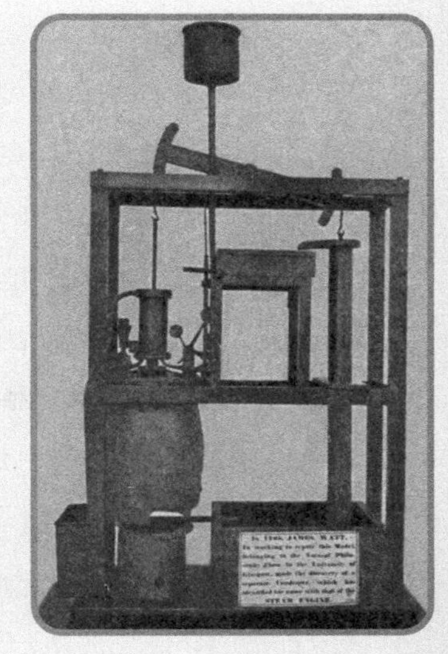

James Watt was experimenting on an older-model steam engine like this one when he figured out how to make a version that wasted less heat.

Now, instead of growing crops and making handicrafts at home, more Americans went to work in city factories where goods were made by machines.

Factories could produce items more quickly and on a larger scale. This led to big changes in how people took care of the land—or didn't.

Soon after European colonizers arrived in the Americas, they started building dams on rivers and streams to power mills that ground grain into flour. Those dams changed the flow of the water, flooding some areas and blocking fish from passing in others. Even more dams were built during the Industrial Revolution. Companies were determined to harness the power of rivers to produce goods, even if that meant the fish couldn't swim their usual routes. Some of America's early environmentalists spoke up about this issue.

"Salmon, Shad, and Alewives were formerly abundant here, and taken in weirs by the Indians, who taught this method to the whites, by whom they were used as food and as manure, until the dam . . . and the factories at Lowell, put an end to their migrations hitherward. . . . Poor shad! Where is thy redress?"

—HENRY DAVID THOREAU, *A WEEK ON THE CONCORD AND MERRIMACK RIVERS,* 1849

In other words: How can a bunch of fish fight back against these big factories?

Henry David Thoreau was an American writer and thinker who believed the United States was too focused on producing goods and making money. He thought spending time in nature was a way to escape that busy buy-and-sell mindset, so he moved to a little house in the woods on Walden Pond in Massachusetts.

The land was owned by his friend Ralph Waldo Emerson, who shared many of Thoreau's ideas. They were known as transcendentalists, people who believed in the powers of the individual and nature and rejected much of organized society.

Ralph Waldo Emerson was a transcendentalist philosopher, writer, and lecturer.

ALONE IN THE WOODS (EXCEPT ON LAUNDRY DAY)

Henry David Thoreau's most famous book, *Walden,* is all about living his life simply in the woods. It's required reading in many schools, but this image of Thoreau tucked away in the forest for two years with nothing

but trees to keep him company needs a little smashing.

The truth is, Thoreau's cabin in the woods was only about a twenty-minute walk from his mom's house in Concord. He would walk into town a few times a week to have dinner and see friends. His mom supported his "simple life in the woods" by sending him home with sandwiches and cookies. Word is, he'd drop off his laundry with her, too.

Both Thoreau and Emerson published essays about their worldviews. Some people paid attention, but most Americans were just excited to see the economy taking off. They weren't interested in protecting the environment if it meant slowing down the nation's growth.

Advances in transportation and energy in the nineteenth century made it even faster for people to cut down forests, clear land, and mine resources. The growing nation needed lumber to build houses, railroads, and ships. In less than eighty years, nearly all Michigan's white pine trees were gone. The rush to log forests in the Pacific Northwest led to habitat loss for bears, wolves, bald eagles, and ravens.

And when gold was discovered in California, thousands of people poured into the state, leaving their mark on the land as well. The search for treasure disturbed riverbeds and damaged habitats for birds, fish, and other wildlife. By 1878, half of California's salmon habitat was destroyed.

Seeking gold in a California river bottom, *Harper's Magazine*, 1860

European settlers in Florida were also making dramatic changes to their landscape. When engineers devised a way to pump water out of wetlands, they started to drain the Everglades—a four-thousand-square-mile watery ecosystem—in hopes of creating more farmland. They ignored the fact that the Everglades were an essential habitat for wildlife. Less than fifteen years after that draining project began, the region's wading-bird population had dropped by 90 percent.

While these natural spaces were being destroyed, American transportation and industry were expanding more quickly than ever. The 1825 completion of the Erie Canal made it easier to transport goods and crops between the East Coast and the Midwest.

The Marriage of the Waters by C.Y. Turner illustrates a scene from the opening of the Erie Canal, 1905.

Railroads were growing as well, and all those trains needed energy to operate. Locomotive engines burned lumber at first, then switched over to coal.

WHAT IS COAL?

Coal is a fossil fuel. It contains energy from plants that lived millions of years ago. When those plants fell into warm, swampy waters, they ended up covered in mud. Layers of plants and mud eventually got

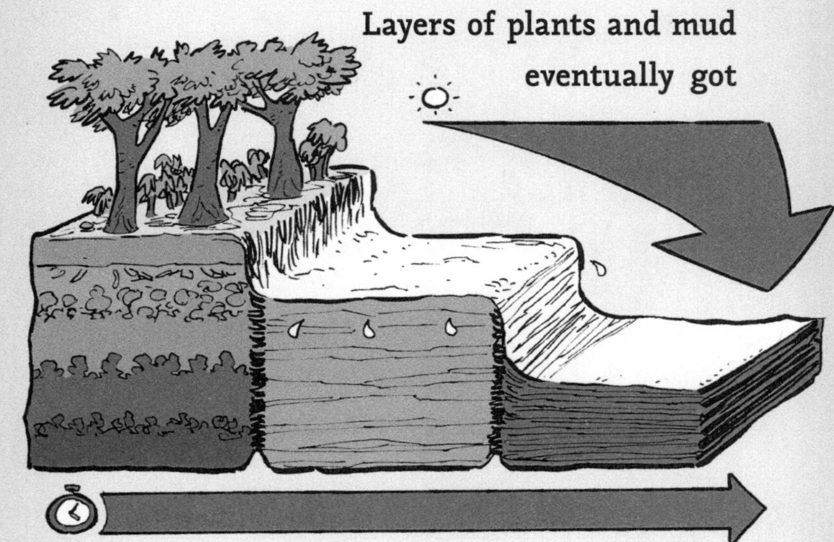

pressed into a substance called peat. As millions of years passed, that peat was buried and subjected to lots of pressure and heat. This combination led to chemical changes that eventually turned the peat into coal.

Nearly all the coal on Earth was produced hundreds of millions of years ago. It's a nonrenewable resource; when it runs out, there won't be any more.

Coal is mostly made up of carbon, which makes it easy to burn for fuel. When coal is burned, it releases carbon dioxide, which is now known to be a cause of global warming. Burning coal also gives off other chemicals and soot, leading to poor air quality and health issues for people.

Coal didn't just provide fuel for railroads; it powered factories and heated people's homes. By the late 1800s, coal was supplying more than half of the energy in the United States, literally fueling the nation's growth.

But it didn't take long for people to notice all that industry was creating pollution. Artists and poets were among the first to speak up. Thomas Cole was a painter and a poet; he used both visual art and words to argue for the preservation of wilderness.

Thomas Cole's 1836 painting *The Oxbow* shows the contrast between wild lands on the left and civilization on the right.

> *"Our doom is near: behold from east to west.*
> *The skies are darkened by ascending smoke."*
>
> —THOMAS COLE, "LAMENT OF THE FOREST," 1838

Sooty skies weren't the only problem. All kinds of industries were expanding in the 1800s. Factories produced goods, and slaughterhouses processed animals into meats. Mills made cloth and paper, while foundries melted iron and cast it into different shapes. All those industries used natural resources and produced tons of waste. Before long, rivers and streams were full of chemicals and animal parts, mixed with other kinds of waste. That included untreated sewage from everyone who lived in the cities and worked in those factories.

In the late 1800s, the factories of New Britain, Connecticut, employed more than three thousand people, and they all had to go to the bathroom somewhere. The privies, or toilets, from those factories drained directly into nearby Piper's Brook. We're talking about two thousand pounds of poop and more than three thousand gallons of urine dumped into the water every single day. Can you imagine the smell? Farmers downstream complained about the stench. Fish were dying, and cattle couldn't drink from the brook anymore.

Just in case you're not already grossed out, you should know those stinky rivers and streams were often the source of drinking water for people who lived downstream from the cities. That wasn't just disgusting; it was deadly. Polluted water led to

epidemics of diseases such as cholera, typhoid, and dysentery.

People who could afford to move started leaving cities to escape the filthy conditions, but not everyone had that option. White people who fled the cities set up laws that made it harder for others to follow. It became difficult or impossible for many Black families and immigrants to buy homes away from the polluted regions. In some communities there were rules against selling homes to people who weren't white. As a result, poor workers and immigrants ended up being blamed for many of the problems they couldn't escape.

This segregation also meant that poor residents were stuck with sooty air, polluted water, and filthy streets. It wasn't fair, and people began to speak up about it. Some of the nation's early environmental movements were aimed at making life better for workers in cities.

CHANGEMAKERS'
YEARBOOK

MOST LIKELY TO USE THE DECLARATION OF INDEPENDENCE TO GET HIS WAY

Harvard Medical School professor **HENRY INGERSOLL BOWDITCH** chaired the Massachusetts Board of Health. He argued that the Declaration of Independence had promised Americans "life, liberty, and the pursuit of happiness." That *had* to include the right to breathe clean air and drink clean water, right?

Thanks in part to his work, Massachusetts lawmakers passed an 1878 law to keep untreated industrial waste and sewage from being dumped into rivers and streams.

THE SAFER-SEWAGE AWARD

ELLEN SWALLOW RICHARDS believed clean drinking water was a basic human right. She was the first woman to enroll at the Massachusetts Institute of Technology (MIT), where later, as an instructor, she led a team of scientists who studied water quality all across the state. Their work led to the first state water-quality standards and the first modern sewage treatment plant, in Lowell, Massachusetts.

THE IF-YOU-WANT-A-JOB-DONE-RIGHT-YOU-HAVE-TO-DO-IT-YOURSELF AWARD

In 1889, **JANE ADDAMS** worked with Ellen Gates Starr to found Hull-House, a settlement house to help immigrants in the growing city of Chicago. People on the city's West Side were struggling with crowded housing and filthy streets, so the women of Hull-House organized a campaign to push for better garbage collection.

Addams thought the city was doing such a crummy job that she put in a bid to collect the garbage herself. The city rejected that bid, but the mayor ended up appointing her local garbage inspector. Addams made so much noise in her new position that the city eventually had to take action and change the way it got rid of the rubbish. (It was kind of embarrassing to be doing such a bad job at garbage collection that a bunch of local citizens had to take over for you.)

THE WALK-IN-THE-PARK AWARD

The pollution in cities led some activists to push for green spaces—wild areas set aside so people who lived in these crowded areas could still enjoy nature.

In 1844, **WILLIAM CULLEN BRYANT**, a poet who was also editor of the *New York Evening Post,* suggested setting aside a large parcel of land in New York City to create "a great municipal park." The site he pitched didn't end up being the final location of New York's Central Park, but his suggestion may have helped put the process in motion.

It was landscape architect **FREDERICK LAW OLMSTED** who ended up designing New York City's grand park, along with Calvert Vaux. The two won a design competition in 1858 and started work on the park right away. Vaux was actually the more experienced of the two designers, but Olmsted had a bigger personality and more social connections, so he ended up getting most of the credit.

In order to create the park, the city earmarked the land they wanted, and people who lived there were forced to give it up. Homeowners were paid, but many argued it wasn't nearly enough. Central Park ended up displacing more than a thousand people, including those who lived in Seneca Village, a thriving community mostly made up of Black people and Irish immigrants.

To add insult to injury, that park, which was supposed to be open to everyone, was actually quite exclusive when it first opened. The park's roads and walking paths were crowded with wealthy people who lived in expensive homes nearby or who could visit in their fancy carriages. Poor working-class people who lived far away couldn't afford the train fare, and the city enacted other rules to limit park use as well.

In the late 1800s and early 1900s, that began to change. The park started holding concerts on Sundays, when working people could attend, and built a playground that was popular with children of middle-class and working-class families. Today, Central Park sees more than forty million visitors a year, from all walks of life.

Across the country, a California botanist named **KATE SESSIONS** made it her mission to ensure that the people of San Diego had access to natural spaces. She made a deal with the city to lease her some land in Balboa Park to create a nursery. In return, she promised to plant a hundred trees in the park each year for ten years, along with hundreds more around the city. Sessions introduced dozens of trees and plants to the city and became known as the Mother of Balboa Park. (Today, ecologists know it's not a good idea to introduce non-native plants to an environment, but that wasn't understood in the 1800s.)

Three
SAVING WILD PLACES

As the turn of the century approached, it became clear that different groups of environmentalists had different beliefs about the relationship between humans and the natural world. They also had different goals.

Activists wanted to focus on protecting people from the conditions that made cities dirty and dangerous places to live. Today, we'd say those people were fighting for environmental justice.

Conservationists wanted to protect nature so it could be used and enjoyed by people. They supported laws that focused on how people could best use land and resources for logging, fishing, and other activities.

And **preservationists** believed that nature should be protected simply because the natural world has value and is worth saving. They pushed to minimize human impacts on nature and to protect existing lands and wildlife from the rapid growth of cities.

Voices from all three of those groups grew louder in the mid-1800s, arguing that the Industrial Revolution had set the wheels in motion for environmental disaster. A Vermonter named George Perkins Marsh was among the first to sound the alarm about the need for conservation in the United States. In 1864, he wrote a book called *Man and Nature,* which argued that people have a responsibility to care for nature, and if they don't, it could mean the end of civilization.

Man is everywhere a disturbing agent. Wherever he plants his foot, the harmonies of nature are turned to discords.

—GEORGE PERKINS MARSH

Marsh and other environmental thinkers eventually convinced the US government to protect some of the nation's land. One of those early conservation efforts focused on Yosemite Valley in California's Sierra Nevada. The region was home to the Ahwahneechee people for thousands of years before white settlers forced them out during the gold rush. When more settlers arrived, it wasn't long before their descriptions and photographs of the area's stunning landscapes made their way east, and tourists soon followed.

An 1855 sketch of Yosemite Valley by Thomas Ayres

Conservationists raised concerns about how all those settlers and tourists might affect the land. They wrote letters, urging the government to protect Yosemite Valley, and in 1864, President Abraham Lincoln signed a bill called the Yosemite Grant, handing Yosemite Valley and a stand of redwoods called Mariposa Grove over to the state of California. It was the first step in protecting the area.

A PICTURE IS WORTH A THOUSAND WORDS

It's important to remember that all these campaigns took place before communication was fast or easy. Most people in the East had never visited Yosemite Valley, so it was hard for them to picture what they were being asked to protect. But that protection still depended on their support. That's why writers and photographers were essential in early pushes for conservation. The same photos and passages that drew tourists to the area could also help people see that it was a place worth fighting for.

Photographer
Carleton Watkins

Carleton Watkins snapped some of those influential photos. He moved from New York to California during the gold rush, and when his pans for gold came up empty, he found success as a photographer. Watkins returned to New York with stunning photographs of Yosemite Valley, which helped persuade the government to protect it.

View from Inspiration Point,
Yosemite Valley, 1879

Yosemite Falls, 2,634 feet, 1876

Section of the Grizzly Giant, Mariposa Grove, Yosemite, 1861

Piwyac, the Vernal Fall, 300 feet, Yosemite, 1861

While the 1864 law protected Yosemite Valley, it wasn't the first official national park in the United States. That honor goes to Yellowstone, which in 1872 became a national park under President Ulysses S. Grant.

Grand Canyon of the Yellowstone
by Thomas Moran, 1872

Yosemite became a national park in 1890, when it was signed into law along with Sequoia and General Grant National Park, which would later become part of Kings Canyon. Other national parks and national monuments soon followed, and states began to set aside land as well. In 1892, New York State created

Today, the Adirondack Park is a mix of private land, where people live and work, and protected public lands that remain "forever wild." Each year, as many as twelve million people visit the Adirondack Park to hike, paddle, bike, camp, and enjoy the outdoors.

the Adirondack Park, one of the nation's first "forever wild" forest preserves.

You might think everyone would welcome these decisions to protect nature. After all, who doesn't love fresh air? But the truth is, the creation of national and state parks wasn't without controversy. Most of us like parks, but how would you feel if the government forced you to give up your home to make space for one? People who lived near protected lands were often treated unfairly, especially if they were poor, Native, Black, or Latine.

The US government forced the Ahwahneechee from Yosemite Valley, even though it had been their home for thousands of years. After the national park was created, some Ahwahneechee people were allowed to stay, living in a village where they sold Native crafts and took photos with park visitors. When the park relocated the "Indian village" and made it smaller, many of the remaining Ahwahneechee lost their jobs and moved away for good. New laws against hunting, fishing, and harvesting wood on public lands suddenly made it a crime for Native people to engage in traditional practices.

For centuries, Shoshone, Bannock, Crow, and Blackfeet people lived on the land that became Yellowstone National Park. With the creation of the park, they were forced onto reservations and denied access to the places where they used to hunt, fish, and harvest trees. White tourists were invited to enjoy that land instead. And European settlers, who hadn't been forced to relocate, kept taking game anyway. Poaching, or illegal hunting, became a huge problem.

MARCH 1894
ONE MORNING IN 1894, A YELLOWSTONE PATROL WAS CHASING DOWN A WELL-KNOWN POACHER NAMED EDGAR HOWELL, WHEN THE OFFICER FOUND A BUNCH OF BISON HEADS HIDDEN IN THE TREES.

NOT LONG AFTER THAT, HE HEARD GUNSHOTS.
POP!
POP!

FROM A DISTANCE, HE SPOTTED HOWELL SKINNING A DEAD BISON.
BUT THE PATROLMAN WAS OUTGUNNED. HIS ONLY HOPE OF CATCHING THIS GUY WAS TO SURPRISE HIM.

SO HE RACED OVER THE SNOW . . .

. . . LEAPING OVER A TEN-FOOT-WIDE DITCH.

The story of Howell's arrest ran in *Forest and Stream* magazine, along with photos of dead bison, captioned "The Butcher's Work." Many people were outraged.

The editor of *Forest and Stream,* George Bird Grinnell, was an early conservationist who wanted

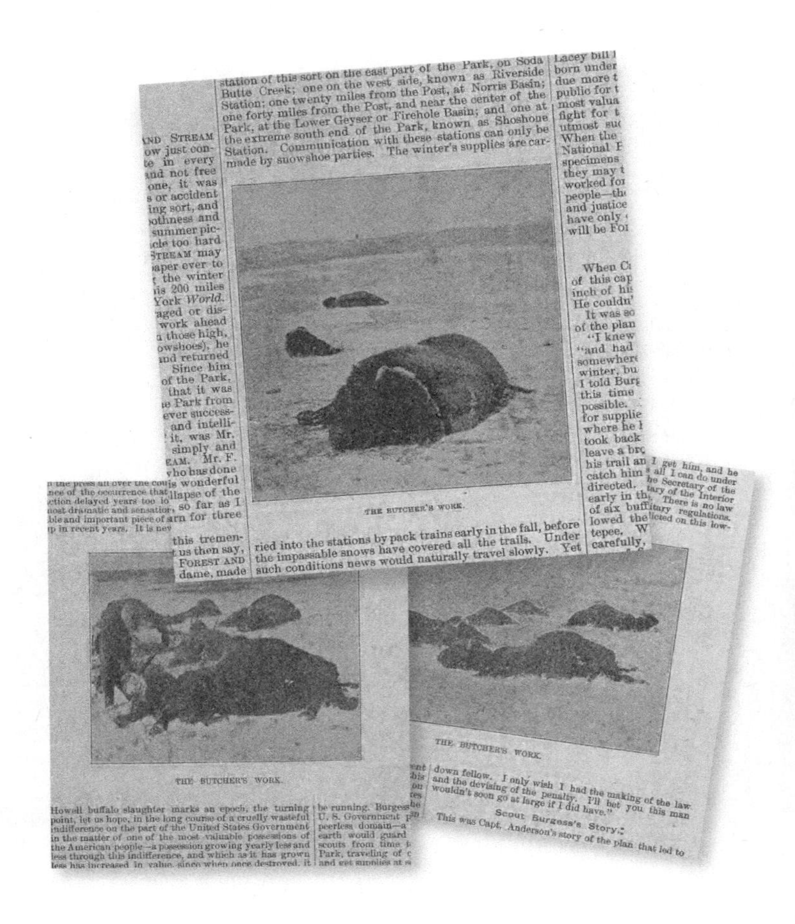
station of this sort on the east part of the Park, on Soda Butte Creek; one on the west side, known as Riverside Station; one twenty miles from the Post, and near the center of the Park, at the Lower Geyser or Firehole Basin; and one at the extreme south end of the Park, known as Shoshone Station. Communication with these stations can only be made by snowshoe parties. The winter's supplies are car-

THE BUTCHER'S WORK.

ried into the stations by pack trains early in the fall, before the impassable snows have covered all the trails. Under such conditions news would naturally travel slowly. Yet

THE BUTCHER'S WORK.

Howell buffalo slaughter marks an epoch, the turning point, let us hope, in the long course of a cruelly wasteful indifference on the part of the United States Government in the matter of one of the most valuable possessions of the American people—a possession growing yearly less and less through this indifference, and which as it has grown less has increased in value, since when once destroyed, it

THE BUTCHER'S WORK.

down fellow. I only wish I had the making of the law and the devising of the penalty. I'll bet you this man wouldn't soon go at large if I did have."

Scout Burgess's Story.

This was Capt. Anderson's story of the plan that led to

tougher laws to protect wildlife in national parks. Soon after that story ran, the US Congress passed a law called the Lacey Act, which made it easier to punish poachers.

BOGUS BISON?

There's just one problem with the dead bison photos in that article. They weren't the bison that Edgar Howell had killed.

The photos in the magazine were from a Smithsonian Institution report that had run seven years earlier. We don't know why Grinnell decided to use the old photos. It's possible he didn't want to wait for the recent photos of the slain bison to be developed.

The real image of Howell's bison heads below probably would have done just as good a job convincing people that park wildlife should be protected.

Poached bison in Yellowstone, 1894

Grinnell was good friends with another conservationist who would go on to sign many more laws aimed at protecting the environment: Theodore Roosevelt.

Even as a kid growing up in New York City, Roosevelt loved wildlife. He devoured books about nature, collected specimens of different animals, and kept live mice in his shirt drawer. (Really!)

As a young man, Roosevelt was elected to the New York State Legislature. But after his wife and mother died of different illnesses on the same day, he took a break from politics and moved to the Dakota Territories to hunt and ranch. There he saw firsthand how European settlers had devastated the bison population. Roosevelt returned east a few years later and teamed up with Grinnell to found a conservation group called the Boone and Crockett Club.

Roosevelt's love of nature continued after he was elected vice president. In fact, he was out hiking in the Adirondacks when a messenger found him to tell him that President William McKinley, who had been shot in Buffalo the week before, had taken a turn for the worse. By the time Roosevelt arrived in Buffalo, the president had died, and Roosevelt was sworn in to take over the job.

Theodore Roosevelt, 1904

As president, Roosevelt greatly expanded the nation's public lands. He established five national parks as well as 150 national forests and eighteen national monuments. Roosevelt also organized the first Conservation Conference in the United States. Governors from across the country joined representatives from Canada and Mexico at the White House in 1908 to talk about how human health depends on the conservation of nature.

Sounds like great work, right? But it's worth noting that when Roosevelt and Grinnell and other Boone and Crockett Club members talked about protecting nature, they were mostly talking about preserving it for wealthy white people. During his presidency, Roosevelt continued earlier policies of removing Native people from their ancestral lands to clear the way for national parks. He didn't see a problem with this action, because he believed white people were superior to both Native people and those of African descent. Roosevelt even wrote a letter to a lawmaker friend, explaining why he didn't think Black Americans in the South should have the right to vote.

—THEODORE ROOSEVELT, FROM A LETTER TO SENATOR HENRY CABOT LODGE, DECEMBER 2, 1916

Roosevelt's racist beliefs—and a growing awareness of them—led New York's Museum of Natural History to remove a statue of him from its grounds in 2022. The statue showed Roosevelt on a horse, with two shirtless men below him, one Native and one African. Roosevelt's great-grandson supported the statue's removal.

"If we wish to live in harmony and equality with people of other races, we should not maintain paternalistic statues that depict Native Americans and African Americans in subordinate roles. The statue of Theodore Roosevelt, my great-grandfather, in front of New York's Museum of Natural History, does so, and it is good that it is being taken down."

—Mark Roosevelt, great-grandson of Theodore Roosevelt

Many of Roosevelt's friends in the conservation movement also held his racist views. Madison Grant, another member of the Boone and Crockett Club, supported national parks, founded the Wildlife

Conservation Society, and oversaw construction of the Bronx Zoo in New York City. He also wrote a book about white supremacy, which included a reference to white people as the "master race." The text went on for more than four hundred pages about how Black people, Native people, Jewish people, and immigrants from many nations were all inferior. (By the way, don't bother looking for information about Grant on the Bronx Zoo or Wildlife Conservation Society websites. Apparently, people there have figured out he wasn't such a great guy, and his name is nowhere to be found.)

You might be wondering why a book about the environment is spending so much time talking about the racism of these conservationists. Does it really matter what they believed, as long as they worked to protect the planet?

It's true that men like Roosevelt had a big impact on environmental policies. It's also true that their personal beliefs, combined with their power, did a lot of harm to some groups of people. And the racism that was part of the early conservation movement didn't just disappear. It would shape the future of environmental issues as well.

THE WILD-MAN AWARD

JOHN MUIR was a Scottish immigrant who moved to Yosemite in 1868 and spent the next forty years hiking and writing about the mountains. Once, on a hike in the Rockies, he reportedly climbed the tallest pine tree all the way to the top and clung there, howling while the wind raged around him.

Muir believed Yosemite was a sacred place that needed protection. His writings helped lead to the creation of Yosemite National Park (1890) and the Sierra Club (1892), a conservation organization that first worked to protect the Sierra Nevada range in California and then expanded to promote environmental causes across the United States.

In the early 1900s, the city of San Francisco experienced a severe water shortage. Officials wanted to build a dam on Yosemite's Tuolumne River, flooding Hetch Hetchy Valley to provide drinking water and hydropower. Muir fought that project with every ounce of his being.

> *These temple destroyers, devotees of raging commercialism, seem to have a perfect contempt for Nature, and instead of lifting their eyes to the God of the mountains, lift them to the Almighty Dollar. Dam Hetch Hetchy! As well dam for water tanks the people's cathedrals and churches, for no holier temple has ever been consecrated by the heart of man.*

—JOHN MUIR, *THE YOSEMITE*

The fight raged on for years, but in 1913, Congress passed the bill and San Francisco got its reservoir. Even though Muir and other preservationists lost this battle, they succeeded in getting a lot more people talking about nature and visiting national parks.

MOST LIKELY TO WRITE DOWN THE DETAILS

ALDO LEOPOLD was a scientist, teacher, and writer known as the Father of Wildlife Ecology. He kept journals about the natural world, and his book *A Sand County Almanac* became an important text in the modern conservation movement. Leopold argued that the environment didn't *belong* to people. Instead, they shared it with every living thing and had a special responsibility to care for it. In 1935, Leopold helped found the Wilderness Society, which worked to expand the nation's wilderness areas.

MOST LIKELY TO ARGUE FOR ALLIGATORS

When the government began draining the Florida Everglades, **MARJORY STONEMAN DOUGLAS** stepped up to defend the ecosystem, an essential habitat for everything from alligators and anhingas to Florida panthers and white-tailed deer. Her book *The Everglades: River of Grass* (1947) became a bestseller and helped people understand that the Everglades were an important natural resource that should be protected. Her work helped lead to the creation of Everglades National Park in 1947.

MOST LIKELY TO RUFFLE FEATHERS

HARRIET HEMENWAY was from a wealthy family and had always worn fancy hats with feathers—until she read about plume hunters who raided a Florida heron rookery. She was horrified to learn that dead birds were left to rot and chicks were orphaned, all so she could have some feathers on her hat. So in 1896, Hemenway teamed up with her cousin and held a meeting to organize with other women and help the birds. Together, they created the Massachusetts Audubon Society, which aimed to protect birds by convincing people not to buy or wear hats with feathers anymore. Instead, they sported "Audubonnets," hats decorated with ribbons.

The Audubon Society became a national group in 1905, and eight years later, Congress passed a law to protect migratory birds.

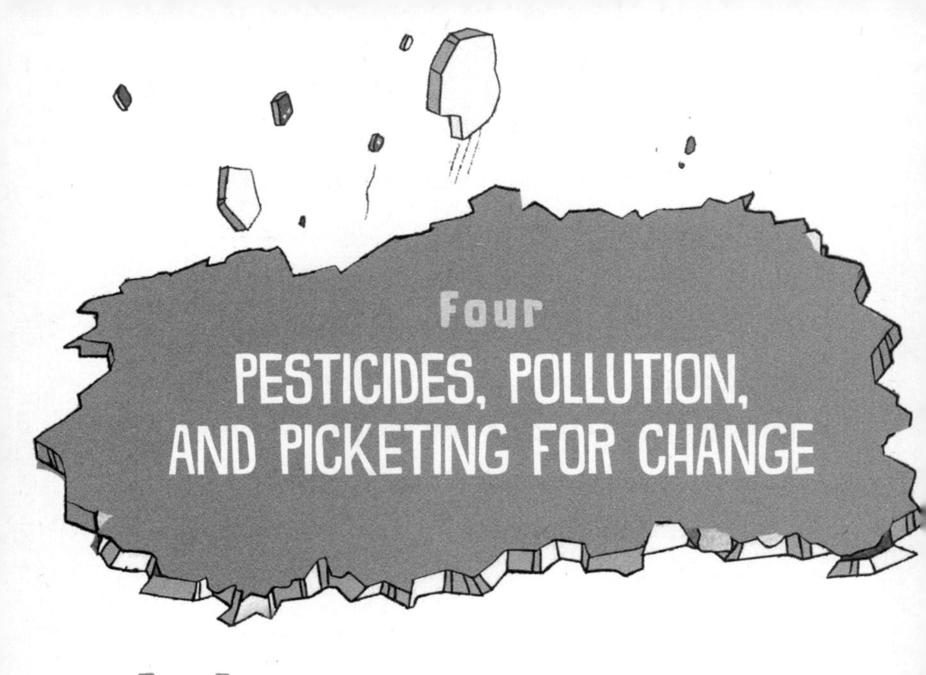

Four
PESTICIDES, POLLUTION, AND PICKETING FOR CHANGE

While the United States was setting aside land to protect nature and wildlife, new technologies both overseas and at home were posing new environmental challenges. Inventions such as cars and electrical appliances meant the nation was burning more coal and oil. Power plants and steel mills were also pumping pollutants into the atmosphere.

In October 1948, the town of Donora, Pennsylvania, was engulfed in a smog so thick and toxic that twenty people died and another six thousand got sick before the weather changed and cleared the foul fog.

Wire mills along the river in Donora, Pennsylvania, filled the sky with smoke in the decades leading up to the deadly 1948 fog. (1910)

Air pollution was also damaging crops in California. Something had to be done.

The Air Pollution Control Act of 1955 set aside money for research about air pollution but left the job of fixing it up to the states. For now, the focus was on learning about the problem rather than taking real action, but this was the first federal law about air pollution.

The nation was also facing environmental challenges as a result of World War II, which took place from 1939 to 1945. Battles around the world forced

US Marines at Guadalcanal in the South Pacific, 1942

American soldiers to travel overseas, where enemy troops weren't the only danger.

Soldiers often had to contend with lice (which carried a disease called typhus) and mosquitoes (which carried malaria). Desperate to defeat the bugs, the United States poured time and money into developing pesticides—chemicals designed to kill disease-causing insects. Early in the war, the US military sprayed oil over larvae sites so baby mosquitoes couldn't develop into biting adults.

But troops had to go where the battles were, and sometimes they moved into new areas that were already thick with adult mosquitoes. This led to the invention of aerosol cans for bug spray—the kind many people still use when they're hiking or camping in the woods.

Originally, those spray cans held pyrethrum, a natural pesticide made from chrysanthemum flowers overseas. But because of the war, there was a shortage, so the United States turned to another chemical: dichlorodiphenyltrichloroethane. (Try saying that three times fast!) Fortunately, it has another name: DDT.

DDT was actually invented in the late 1800s, but it wasn't until World War II that the government discovered the chemical could kill lice and mosquitoes, even in low doses. It worked longer than pyrethrum did, too.

The US government was eager to protect its soldiers right away, so safety testing happened quickly, and soon pilots were spraying DDT from airplanes in the South Pacific. It wiped out mosquitoes and lice like magic and even stopped a typhus outbreak in Naples, Italy, in 1943.

People called it a miracle poison. They designed posters and wrote poems about it.

This World War II–era poster urged people to use DDT to protect their victory gardens, which many Americans planted as a way to help with the war effort at home.

Little insect, roach, or flea,
Have you met with DDT?
In the foxhole, up the line,
DDT gets eight in nine.
In the tank, beside the gun,
DDT means battles won.

This chemical company advertisement for DDT appeared in *Time* magazine in 1947.

But DDT didn't turn out to be such a miracle. Insects started to develop resistance. Populations of mosquitoes that carry malaria began to bounce back. On top of that, scientists discovered that the chemical had a devastating impact on wildlife. A marine biologist named Rachel Carson made it her mission to bring attention to the destruction DDT caused.

RACHEL CARSON LOVED WILDLIFE, EVEN WHEN SHE WAS A LITTLE GIRL.
SHE EARNED A MASTER'S DEGREE IN ZOOLOGY AND THEN WORKED WITH THE US FISH AND WILDLIFE SERVICE FOR FIFTEEN YEARS.
1951
CARSON WROTE A BOOK ABOUT OCEAN WILDLIFE.
THE SEA
RACHEL L. CARSON
THEN SHE TURNED HER ATTENTION TO PESTICIDES LIKE DDT.

BY THE 1950S, THE UNITED STATES WAS SPRAYING MILLIONS OF POUNDS OF DDT EACH YEAR . . .
DDT
. . . TO WIPE OUT EVERYTHING FROM MOSQUITOES TO FIRE ANTS.

PEOPLE SPRAYED DDT ON THEIR DISHES AT HOME.

THEY EVEN PUT DDT-TREATED WALLPAPER IN THEIR KIDS' BEDROOMS.

The newly hatched chick is clothed in a soft gray down; in only a few hours it takes to the water and rides on the back of the father or mother.

CARSON DESCRIBED THE IMPACT DDT HAD ON THE WESTERN GREBES THAT LIVED AROUND CALIFORNIA'S CLEAR LAKE.
AFTER DDT WAS SPRAYED TO KILL GNATS IN THE AREA, THOSE CUTE BABY BIRDS BEGAN TO DIE.

OTHERS NEVER EVEN GOT A CHANCE TO HATCH.
DDT CHANGED THE WAY THE BIRDS' BODIES PROCESSED CALCIUM, RESULTING IN THINNER EGGSHELLS THAT BROKE WHEN A PARENT SAT ON THEM TO KEEP THEM WARM.

But people didn't ignore *Silent Spring*. Rachel Carson was invited to testify before Congress about the dangers of DDT.

Carson worked with other environmental activists who used public education, legislative hearings, and letters to Congress and newspapers to raise awareness of DDT's negative effects. They got the attention of President John F. Kennedy, who appointed a special scientific advisory committee to study DDT.

Rachel Carson and wildlife artist Bob Hines researching in Florida

In 1972, the pesticide was banned for most uses in the United States. Four years later, Congress passed another law, called the Toxic Substances Control Act, which created additional rules about the use of many other chemicals.

Carson and others who spoke up did more than just raise awareness about DDT. They inspired people to see themselves as environmentalists—citizens who could take action to help the environment.

BIRDS AND BRIEFS: THE BEGINNING OF ENVIRONMENTAL LAW

Sometimes, environmental activists fight their battles in court. When birders on

Long Island in New York saw what DDT was doing to local bird populations, they sued the government to stop spraying the chemical. That lawsuit helped pave the way for the federal ban on DDT. It also helped establish the right of regular people to sue their government over environmental issues.

Groups such as the Environmental Defense Fund, Earthjustice, and the Natural

Resources Defense Council use both science and the law in their court battles. Today, the field of environmental law is dedicated to protecting our air, water, and natural resources.

When people learned about the impact DDT had on wildlife, some growers started using different pesticides that didn't last as long. This decision was better for the birds, but some of the new chemicals were even more dangerous to the people applying them. California farmworkers, most of whom were of Mexican descent, were exposed to those harmful toxins, and many became ill.

You might think the environmental groups that had spoken up for wildlife would spring into action to protect these farmworkers, too. But they didn't. Farmworkers said the lack of action amounted to environmental racism. Was their health less important just because they were Mexicans and Mexican Americans?

Many of the groups that had fought for the birds were nowhere to be found. But two labor leaders and civil rights activists did step up to help.

Cesar Chavez

Dolores Huerta

Cesar Chavez (center) marches with United Farm Workers members in Redondo Beach, California, in 1975.

Cesar Chavez and Dolores Huerta founded the National Farm Workers Association. The group later joined forces with the Agricultural Workers Organizing Committee to create the United Farm Workers union. They led thousands of workers to march for higher wages and safer working conditions.

United Farm Workers called pesticides "the poisons we eat" in its materials promoting the 1965 grape boycott.

Marion Moses was a nurse who worked with Chavez to care for farmworkers. She set up a health and safety commission to investigate their exposure to dangerous chemicals. She also promoted a national boycott of table grapes to raise awareness about pesticides. She wanted people to know this wasn't just an issue for farmworkers. Those chemicals were literally on everyone's food.

The farmworkers succeeded in gaining some health and safety protections. Growers agreed to stop using several pesticides and promised to follow rules for the use of other chemicals.

By now, you've probably noticed that some environmental groups run by white leaders were still holding on to racist ideas about who deserved protection and who didn't. Some groups didn't even allow people of color to be members. The Prairie Club, which was a midwestern version of the Sierra Club, was officially "open to white people of any nationality or creed."

Shut out by white environmentalists, Black and Indigenous people and other people of color started their own groups. And they got things done. In the 1960s and 1970s, Black activists in St. Louis set up blood tests for thousands of children to find out if lead paint was making them sick. When they determined that it was, they pushed lawmakers to pass and enforce laws against using the toxic paint.

More and more citizens were coming to understand that pollution was affecting public health. And when enough people speak up, lawmakers are forced to listen.

Five

CLEANING UP OUR ACT: THE FIRST EARTH DAY CELEBRATION

Public interest in the environment took off in the 1960s. Groups like the National Audubon Society and the Sierra Club saw an explosion in membership. In 1962, President John F. Kennedy held a White House Conference on Conservation.

—FROM JOHN F. KENNEDY'S REMARKS TO THE WHITE HOUSE CONFERENCE ON CONSERVATION

Kennedy proposed a Land and Water Conservation Fund, using money from offshore oil drilling to buy land for national parks and recreation areas. He also supported the Clean Air Act of 1963 to address air pollution but was assassinated before it became law. Kennedy's successor, Lyndon Baines Johnson, signed the Clean Air Act as well as the Wilderness Act, which established a national preservation system that now protects more than eight hundred wildlife areas.

Environmental awareness was definitely on the rise, but the last year of the decade brought several disasters that made the issue nearly impossible to ignore. In 1969, a blowout at an oil rig off the California coast caused the largest oil spill of its time in American waters. Millions of gallons of oil spilled into the Santa Barbara Channel, blackening beaches and killing thousands of birds and marine mammals. Volunteers scrambled to save affected wildlife.

Later that year, another dramatic image made headlines—a polluted Ohio river caught on fire. The Cuyahoga River, which runs through Cleveland, was full of litter and slicked with oil. All it took was a spark—likely caused by a passing train—and the river went up in flames.

It had actually caught fire a number of times before, but this incident ignited the nation's attention. The fire wasn't that big; firefighters put out the flames in less than thirty minutes, but the blaze still sparked concern and activism. How could a river be so polluted it actually caught *fire*?

FANNING THE FLAMES

A *Time* magazine article on the Cuyahoga River fire helped to fuel outrage over the polluted waters. But the photo used in that piece didn't match the story. When reporting on the 1969 river fire, *Time* featured a

Cuyahoga River fire, Jefferson Avenue and West Third Street, 1952

photograph of a 1952 fire, which was much bigger.

There are no known photos of the 1969 fire, but when the article ran, few people cared that this picture was outdated. They understood that pollution was a problem, and something had to be done.

Even the air was dangerous. That July, California radio and TV stations announced that smog in some cities had made it unsafe for kids to play outside.

These environmental messes inspired people to come together and raise their voices. And now Americans knew how to do that. The 1960s had already been filled with protests and demonstrations as people spoke up about all sorts of things—from civil rights to women's rights to the Vietnam War. Now they'd put their voices to work for the earth.

HE WANTED LAWMAKERS IN WASHINGTON, DC, TO DO MORE TO HELP THE ENVIRONMENT.
AND HE HAD AN IDEA FOR HOW TO GET THEIR ATTENTION.
SEATTLE, SEPTEMBER 1969
I AM CONVINCED THAT THE SAME CONCERN THE YOUTH OF THIS NATION TOOK IN CHANGING THIS NATION'S PRIORITIES ON THE WAR IN VIETNAM AND ON CIVIL RIGHTS CAN BE SHOWN FOR THE PROBLEM OF THE ENVIRONMENT.

THAT IS WHY I PLAN TO SEE TO IT THAT A NATIONAL TEACH-IN IS HELD.
THE TIME WAS RIGHT FOR NELSON'S BIG IDEA.
PEOPLE FLOODED HIS OFFICE WITH PHONE CALLS, ASKING HOW THEY COULD PARTICIPATE.
APRIL 22, 1970
THAT APRIL, NELSON'S VISION BECAME A REALITY. HE ORGANIZED A DEMONSTRATION FOR THE EARTH. ONE TOO BIG FOR POLITICIANS TO IGNORE.
LANET
ARTH
PROTECT THE PLANE
IT'S
TO
LA
UP TO TWENTY MILLION AMERICANS PARTICIPATED.

CROWDS MARCHED ALONG
NEW YORK CITY'S FIFTH AVENUE.

STUDENTS STAGED
TEACH-INS TO HELP PEOPLE
LEARN HOW THEY COULD
PROTECT THE ENVIRONMENT.

THE FIRST EARTH DAY WASN'T JUST
ONE BIG DEMONSTRATION. IT WAS
TWELVE THOUSAND INDEPENDENTLY
RUN EVENTS ALL OVER THE
COUNTRY. FROM CALIFORNIA . . .

. . . TO COLORADO . . .

The first Earth Day was so popular that lawmakers decided they should get involved, too. Congress adjourned for the day so members could go home and participate in Earth Day events in their communities. It wasn't just one political party supporting the environment—it was *everybody*. And Earth Day was only the beginning. That one day of activism planted a seed that would continue to grow.

Six
EARTH DAY ENERGY

In the months following the first Earth Day, many new environmental groups emerged. The members of these groups all cared about protecting the planet. However, they didn't always agree about how to get that job done. Some argued that the path to progress was through new laws, court cases, and the election of environmentally minded leaders. Others staged many kinds of protests, marching, demonstrating, and even physically blocking projects they thought would harm the earth.

THE ENVIRONMENTAL GROUP GREENPEACE GOT ITS START WITH A DOZEN MEN IN AN OLD FISHING BOAT.

GREENPEACE

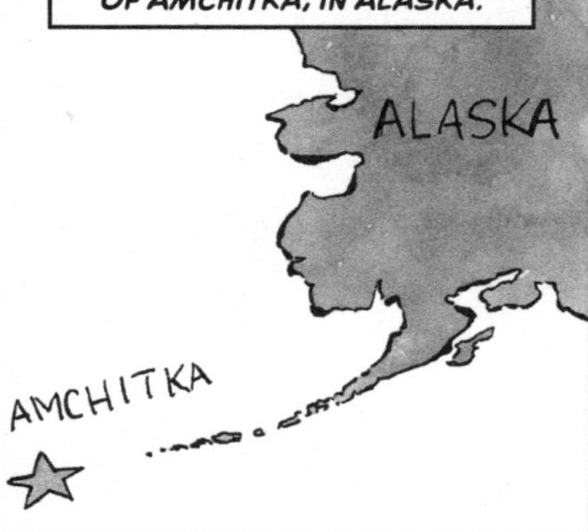

THEY FACED
BRUTAL WEATHER . . .

. . . AND THE US MILITARY
INTERCEPTED THE SHIP
BEFORE IT COULD REACH
THE TESTING SITE.

THE NUCLEAR TEST WENT ON AS PLANNED . . .
. . . BUT GREENPEACE'S FAILED ATTEMPT TO BLOCK IT
GENERATED ENOUGH PUBLICITY THAT THE GOVERNMENT
CANCELED ALL FUTURE TESTS AT THE SITE.

Other groups took a more moderate approach. Environmental Action was an advocacy group devoted to education and lobbying. ("Lobbying" means trying to persuade lawmakers to vote a certain way.) The group put out a list called the Dirty Dozen, naming and shaming politicians who voted against environmental protection policies.

Another organization, the League of Conservation Voters, was formed to keep track of how members of Congress voted on environmental issues. That made it easier for citizens who cared about the environment to vote for lawmakers who shared their concerns—and to vote out those who didn't.

Because voters were paying more attention to the environment in the 1970s, lawmakers paid attention, too. Even before the first Earth Day, President Richard Nixon signed the National Environmental Policy Act into law. It required builders to create an environmental impact statement—a study that outlines how a project could affect the surrounding plants, animals, water, air, and soil—for any project that received federal funding.

Nixon called for the creation of the Environmental Protection Agency, or EPA, to protect the nation's air and water quality and monitor pollutants. His plan also included the formation of a new group called the National Oceanic and Atmospheric Administration (NOAA) for air and sea research. The Clean Air Act of 1970 assigned the EPA to make laws regulating emissions that pollute the air.

President Richard Nixon signs the Clean Air Act of 1970 into law.

In 1972, Congress updated an old water pollution law, giving the EPA authority to set standards for wastewater to keep pollution from pouring into rivers and streams. This revised law became known as the Clean Water Act. It also funded new sewage treatment plants to keep raw sewage from being dumped into waterways. Finally, a solution to the poopy-river problem!

WHAT HAPPENS WHEN YOU FLUSH THE TOILET?

You may not think too much about what happens to your poop and pee after you flush the toilet. But thankfully, other people have figured out how to keep human waste from contaminating our waterways. When you flush, the wastewater likely travels through pipes underground to a sewage treatment plant.

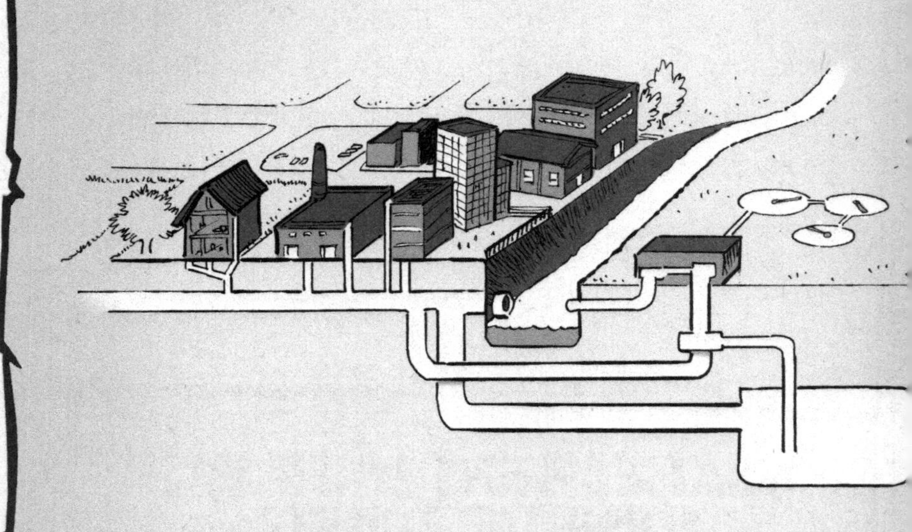

Once there, the wastewater goes through a series of treatments to remove pollutants. First, the water is screened to get rid of large particles. Next, bacteria is introduced to break down the organic matter in the wastewater. Then the treated wastewater is disinfected with chlorine to get rid of any remaining bacteria. Many communities require additional steps to remove pollutants such as nitrogen and phosphorus, which can cause problems in lakes and rivers. Once this process is complete, the cleaned-up wastewater is pumped into a nearby body of water.

THE EARLY 1970S SAW A SLEW OF ENVIRONMENTAL LAWS ENACTED. THE FEDERAL GOVERNMENT BANNED DDT IN 1972.
DDT

THE MARINE MAMMAL PROTECTION ACT ALSO BECAME LAW THAT YEAR, AFTER SCIENTISTS REPORTED THAT POPULATIONS OF SEA LIONS, WHALES, AND OTHER ANIMALS WERE DECLINING BECAUSE OF HUMAN ACTIVITIES.

THE NEW LAW PROTECTED ALL MARINE MAMMALS, FROM DOLPHINS AND MANATEES . . .
. . . TO POLAR BEARS AND WALRUSES.

All around the world, people were paying more attention to the earth. In 1972, the United Nations Conference on the Human Environment took place in Sweden. Attendees focused on four areas of concern: balancing the environment with development, human settlement, natural resources, and pollution. This meeting led to the establishment of the United Nations Environmental Programme, which became the leading global authority on environmental issues.

THE ENDANGERED SPECIES ACT TAKES FLIGHT

Before European settlers showed up, there were millions of passenger pigeons in North America—probably even *billions*. But by the late 1800s, many of the trees they depended on had been chopped down, and the birds themselves had been hunted for cheap meat. The last known wild passenger pigeon was seen in Ohio in 1900. A teen-aged boy mistook it for a different kind of bird and shot it. The last living member of the species—a pigeon named Martha—died at the Cincinnati Zoo in 1914. Her body was donated to the Smithsonian Institution.

It wasn't until 1973 that the United States enacted a federal law to protect all endangered wildlife. The Endangered Spe-cies Act required federal agencies to use

Martha, a passenger pigeon

scientific evidence to make a list of all species in danger of extinction so steps could be taken to help those populations recover. The law also gave the federal government permission to limit land use where endangered plants and animals live.

When that law took effect, there were just a few dozen California condors left in the wild. But unlike the passenger pigeon, the condors got a second chance when people took action. Wildlife biologists raised baby condors, feeding them with hand puppets designed to look like their parents.

They taught the birds not to eat garbage by setting up trash that delivered a mild shock when birds touched it. They vaccinated condors against a disease called West Nile virus and, later on, against avian influenza as well. The huge conservation effort paid off. By 2023, the California condor population had increased to about 560 birds.

A biologist feeds a California condor chick with a hand puppet.

Seven
AFTER EARTH DAY

By now, you might be thinking Earth Day was the turning point that set the world on its way to a cleaner environment. But the truth is, the energy of Earth Day didn't last forever. Public support for environmental issues tends to fade when other challenges arise and make people's lives difficult. And by the mid-1970s, many were struggling.

In 1973, the Middle Eastern nations that produce much of the world's oil announced an embargo, refusing to sell oil to countries that had supported Israel in the recent Yom Kippur War. The United States was one of those nations. Suddenly, Americans were being asked to pay more for gasoline, heating oil, and other products.

Soon after, the United States announced new standards for fuel economy, setting a national fifty-five-mile-per-hour speed limit (driving faster than that burns more fuel) and pushing automakers to build more energy-efficient vehicles. The car companies didn't like that. Why should *they* have to spend more money to make cars that burned less gas? It was hurting their business! How were they supposed to make a profit and provide jobs for people who needed them?

On top of all that, in the middle of the oil crisis, President Nixon was forced to resign. His administration had been involved in a burglary at the Democratic National Committee headquarters, and Nixon had tried to cover it up. (This became known as the Watergate scandal, named for the office building where the break-in occurred.) When Gerald Ford took over as president, he made it clear that he was okay with protecting the environment—but not if it ended up costing jobs.

—FROM GERALD FORD'S REMARKS AT THE NATIONAL ENVIRONMENTAL RESEARCH CENTER IN CINCINNATI, OHIO, ON JULY 3, 1975

Even with less public support, Congress passed a handful of new environmental laws through the mid-1970s.

It was becoming clearer that toxins in the air, water, and ground affected not only wildlife but also humans. In 1978, hundreds of people had to evacuate their homes in the Western New York neighborhood of Love Canal because their houses had been built on a toxic waste site.

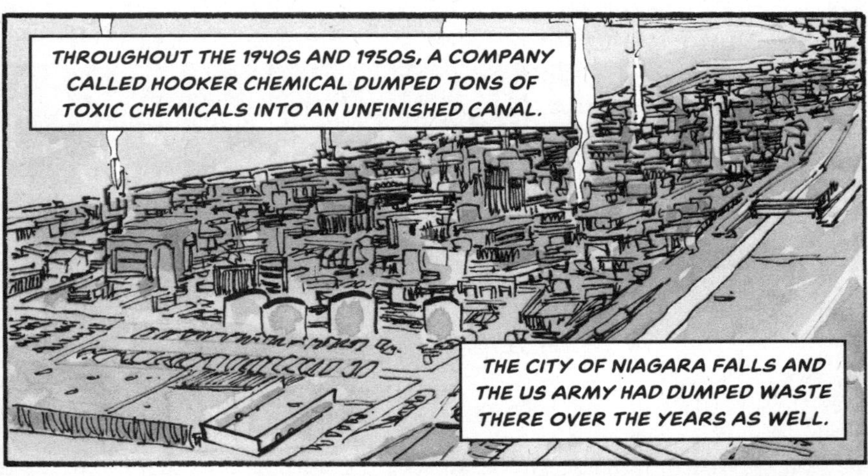

. . . AND SOLD THE LAND TO THE LOCAL SCHOOL BOARD FOR A DOLLAR.
$1
INFORMATION ABOUT WHAT WAS BURIED UNDER ALL THAT SOIL WAS INCLUDED IN THE CONTRACT, BUT NO ONE PAID MUCH ATTENTION AT THE TIME.

THE COMMUNITY BUILT A SCHOOL ON THAT SITE.

PEOPLE BUILT HOMES THERE, AND FAMILIES MOVED IN.

FOR DECADES, PEOPLE LIVED IN THAT NEIGHBORHOOD, WHERE THEIR KIDS ATTENDED SCHOOL AND PLAYED OUTSIDE.

NO ONE KNEW THAT POISONS WERE LEAKING FROM RUSTY BARRELS BURIED IN THE DIRT.
AT LEAST NOT UNTIL THE LATE 1970S, WHEN RESIDENTS OF LOVE CANAL STARTED GETTING SICK.

TOXIC CHEMICALS HAD LEAKED INTO THE SOIL AND THE WATER SUPPLY.

WHEN THE EPA ARRIVED TO INVESTIGATE, SCIENTISTS CONFIRMED THAT THE CONTAMINATION WAS TO BLAME FOR THE ILLNESSES.
SOME OF THE CHEMICALS WERE KNOWN TO CAUSE CANCER.

EPA OFFICIALS WHO VISITED
THE SITE CALLED IT AN
"ENVIRONMENTAL TRAGEDY."

WASTE DISPOSAL DRUMS
HAD SURFACED FROM
UNDER THE SOIL.

TREES AND GARDENS
WERE DYING.

PUDDLES OF POISON BUBBLED UP IN
PEOPLE'S YARDS AND BASEMENTS. EVEN
ON THE GROUNDS OF THE LOCAL SCHOOL.

When a mom named Lois Gibbs tried to transfer her son to a safer school, the school board told her no. So she and her neighbors staged protests and held rallies.

The EPA had already reported that the chemicals seeping out of those barrels could cause long-term damage—not just to the people living on the land now but to their children and grandchildren, too. But the government refused to help people move.

Love Canal neighbors were furious. They invited two EPA officials to one of the abandoned houses to talk with residents about their study. When the EPA officials tried to leave, the residents blocked their path and held them hostage for about six hours, hoping to get President Jimmy Carter's attention.

Eventually, the federal government stepped in, using millions of dollars in emergency funds to move more than nine hundred families out of the area and purchase their polluted properties.

Abandoned land in the Love Canal neighborhood

But not all residents of the Love Canal area were treated equally. Poor Black families living at a nearby public housing project were also worried about their exposure to toxins. They rented their homes instead of owning them, which meant they weren't included in the Love Canal homeowners group. And they weren't getting the same help to move.

So they worked with the National Association for the Advancement of Colored People (NAACP) to form their own group, the Concerned Love Canal Renters Association. Its leaders, Elene Thornton and later Sarah Herbert, argued that since the government had paid to purchase white families' homes, they should help Black renters, too. After more than a year of fighting, the housing project families were finally included in the plan and given the assistance they needed to move from the toxic site.

The disaster at Love Canal spurred Congress to pass a law known as the Comprehensive Environmental Response, Compensation, and Liability Act, or Superfund. It set up a system to pay for cleanups of the worst polluted sites in the nation. Because Love Canal wasn't alone. The EPA estimated there could be up to fifty thousand other toxic waste sites, and about two thousand of them likely posed major health risks.

President Carter signs the Superfund bill into law on December 11, 1980.

Sounds like a positive step, right? But don't start cheering just yet. Remember how car companies fought against laws requiring more fuel-efficient vehicles? Corporations didn't want to be held responsible for their pollution practices, either. They fought the new regulations of the 1970s with all their might. They warned political leaders that too many laws protecting the environment would ruin the nation's economy, keeping industries from making money and being successful.

When another oil crisis sent gas prices soaring in 1979, President Carter suggested a number of solutions. He proposed a tax on cars that guzzled gas, incentives for people to better insulate their homes, and investments in solar and wind energy. He even had solar panels installed on the White House.

But Carter's presidency was also marked by a recession—a time when the economy wasn't doing well. It was hard for many workers to find jobs, and inflation was high, too. That meant everything cost more, and many people couldn't afford the things their families needed. When Ronald Reagan became president in 1980, he promised to fix all those problems.

Reagan made it his mission to roll back environmental rules that businesses didn't like. He cut EPA funding as well as the research budget for renewable energy sources. And remember those solar panels Carter had installed on the White House? Reagan had them taken down while the roof was being redone.

During this period, some Americans started thinking of environmentalists as being against job creation and economic growth. That's when saving the earth became less of an everybody issue and more of a political issue, which is why you might hear people from different political parties arguing about it today.

GOING NUCLEAR

The 1970s also brought debates over nuclear power, which is created when atoms are split apart, releasing energy. That energy heats

up water to produce steam, and that steam turns big turbines and generates electricity.

Nuclear power is a low-carbon energy source—it doesn't contribute as much to global warming (more on that soon)—but there are also drawbacks. Nuclear power produces dangerous radioactive waste, which needs to be disposed of . . . somewhere. Even people who support nuclear power are

often against having those toxic waste sites near their homes.

Nuclear power also carries a risk of serious accidents, like the one that happened at Pennsylvania's Three Mile Island nuclear plant in 1979. When a cooling system failed, one of the plant's reactors partially melted down, releasing dangerous radioactive gases into the environment. The plant shut down, and remarkably, no one was hurt or killed. But tens of thousands of people had to evacuate their homes. Cleanup of the site lasted fourteen years, and the plant was officially shut down in 2019. It was the worst nuclear disaster in US history and drastically changed the safety regulations for nuclear plants. It also made many people more wary of nuclear energy, even though many safety issues have since been addressed.

THE GRANDMOTHER OF CONSERVATION

MARGARET MURIE, who went by Mardy, was known as the Grandmother of the Conservation Movement. She grew up in Alaska, and her husband, **OLAUS**, was a biologist. (Their honeymoon was an eight-month expedition to research caribou!)

Murie spent her life studying and writing about nature. She worked on the Alaska National Interest Lands Conservation Act of 1980, a law that protected more than one hundred million acres of Alaska wilderness.

THE PLANETWALKER AWARD

After witnessing a 1971 oil spill in San Francisco Bay, **JOHN FRANCIS** stopped using motorized transportation and decided to walk across the United States to promote awareness of the environment. For seventeen years, he also didn't speak—he walked all over the United States and South America, journaling and sketching to share his ideas. Later, Francis became a National Geographic Society education fellow and program director for an organization called Planetwalk, which teaches grade-school students about the environment.

THE BEACH LADY AWARD

MAVYNEE BETSCH was a trained musician who toured the world as an opera singer. In 1975, she became a full-time environmentalist working to protect the beach she loved. Her great-grandfather Abraham Lincoln Lewis founded Florida's oldest African American beach, called American Beach, in 1935. At that time, Black people weren't allowed to visit most beaches due to segregation, so American Beach was a treasured getaway for African Americans. When the civil rights movement opened up access to more beaches in the 1960s, American Beach wasn't as busy, and conditions began to decline.

But Betsch believed the beach was an important part of history as well as an essential natural area that should be preserved. In the 1970s, she moved to Florida and made it her mission to protect American Beach from being destroyed or developed. Today, the beach is listed on the National Register of Historic Places.

Eight
TRASH, TREE HUGGERS, AND TOXIC WASTE

Sometimes we talk about the environment as if it's one big-picture issue that affects all people the same way. But not everyone has the same access to clean air, water, and land. Poor people, Native people, Black people, and other people of color have often felt the worst impacts from environmental disasters.

By the late 1970s, growing awareness of this unfair situation led to a movement focused on environmental justice. It's a simple idea, really—that all people have the right to an environment that's clean and safe, no matter where they live, how much money they have, or what their background might be. Seems hard to argue with that, doesn't it? So what created the inequity in the first place?

Early leaders of the conservation movement tended to be wealthy and white, and many held racist beliefs. Rich and middle-class white people also held most positions of power in government. Those in charge rarely approved garbage sites or toxic waste dumps to be located in their own neighborhoods. Instead, such projects were frequently dumped on the doorsteps of poor people and members of historically marginalized groups—Indigenous people, Black people, and other people of color—who did not have a voice standing up for them in the government. Many communities in low-income housing had to deal with nearby garbage dumps and environmental toxins polluting their air, water, and land. It wasn't fair, and local leaders began to speak up about the injustice.

Linda McKeever Bullard was a lawyer who helped her clients sue the state of Texas in 1979. They were fighting a landfill planned for a Black neighborhood in Houston. Linda's husband, Robert Bullard, was a sociologist and environmental activist. He wanted to help with his wife's case, so he led a survey of Houston neighborhoods, looking for evidence of environmental racism. He found that in a city where just one in four residents was Black, *all* of the city-owned landfills and most of the incinerators were located in Black neighborhoods. Houston was literally dumping its garbage on Black communities.

Linda Bullard didn't win the case, but Robert was inspired to do more research and found similar stories playing out all across the South. He wrote books about environmental racism and became co-chair of the National Black Environmental Justice Network, a coalition of groups fighting for similar goals. Bullard came to be known as the Father of Environmental Justice.

Meanwhile, a Chicago woman named Hazel Johnson is often considered the mother of the movement.

Environmental activist
Hazel Johnson

Johnson lived in Altgeld Gardens, a Chicago public housing project where an unusual number of people seemed to be getting sick. After her husband died of cancer in 1969, she learned that her neighborhood had the highest cancer rates in the city. Johnson organized a campaign to file complaints with the state's Environmental Protection Agency. Her efforts yielded a study that showed the housing project was polluted—and dangerous.

Altgeld Gardens was built at the center of what Johnson called a "toxic donut"—a ring of dozens of landfills, sewage-filled waterways, and leaking storage tanks. She founded a group called People for Community Recovery (PCR) and led the fight for safer sewer and water lines for the neighborhood.

A MOUNTAIN OF CONCRETE AND DUST

The idea of environmental justice caught on. People in communities all over the country raised their voices and picked up their pens in the fight for cleaner places to live. In 1993, the city of Huntington Park, California, gave a company permission to open a concrete recycling shop in town. The following year, an earthquake wiped out part of a freeway, and chunks of the road were brought to the shop to be recycled—or at least that was the plan. Instead, a giant heap of concrete towered over the neighborhood for years. The mostly Spanish-speaking residents called it La Montaña—the mountain.

Dust from the concrete heap was everywhere. Neighborhood kids were always coughing, and some residents developed asthma and other breathing problems.

Worried about their long-term health, residents worked with local environmental groups and lobbied the city council to revoke the recycling company's permit and remove the mountain of debris. In 2004, after ten years of fighting, the cleanup finally began. Eight years later, Linda Esperanza Marquez High School was established in Huntington Park. It was named for one of the most vocal leaders in the fight for La Montaña's removal and the safety of the Huntington Park neighborhood.

When George H. W. Bush was elected president in 1988, he called himself the Environmental President. Whether or not he deserved the nickname was up for debate, but during his time in office, the government funded the cleanup of a number of nuclear power plants and toxic military sites. Bush signed the Elwha River Ecosystem and Fisheries Restoration Act, which ordered the removal of a dam in Washington state to help bring back salmon.

The Elwha Dam, shown here, was removed from the river in 2012.

Bush also signed a 1990 law to help prevent oil-spill disasters. This action was in response to the 1989 wreck of the *Exxon Valdez* off the south coast of Alaska, which caused the worst oil spill in history at the time. The tanker hit a reef, sending nearly eleven million gallons of oil surging into Prince William Sound, an essential habitat for wildlife.

The *Exxon Valdez* ran aground in Prince William Sound on March 24, 1989.

The *Exxon Valdez* spill killed more than two hundred thousand seabirds, hundreds of bald eagles, nearly three thousand otters, and twenty-two orcas. People were outraged. They pressured elected leaders to make sure nothing like this could happen again. In response, lawmakers passed the Oil Pollution Act of 1990, which required oil companies to use double-hull tankers to help prevent spills. It also increased penalties for companies responsible for spills, made it easier and faster for the EPA to respond, and set up a fund to pay for cleanups.

Workers clean a shoreline after the *Exxon Valdez* spill.

Western redwood forests served as another environmental battleground. After Pacific Lumber Company started cutting down more old-growth trees in the 1980s, activists organized a three-month campaign called Redwood Summer and used a variety of tactics to slow down logging.

SOME DROVE SPIKES INTO TREES, A PRACTICE THAT WAS DANGEROUS FOR LUMBERERS.
WARNING! THIS TREE HAS BEEN SPIKED. DO NOT USE A CHAIN SAW ON THIS TREE OR YOU RISK BREAKING THE CHAIN SAW.
IN 1987, A SAWMILL WORKER WAS SERIOUSLY INJURED WHEN HE WAS CUTTING A TREE AND HIS BLADE HIT A STEEL SPIKE. THE INCIDENT CAUSED MANY PEOPLE, INCLUDING ENVIRONMENTALISTS, TO SPEAK OUT AGAINST SPIKING TREES.

OTHER ACTIVISTS TOOK UP RESIDENCE IN THE REDWOODS THEY WERE TRYING TO PROTECT.

SOMETIMES THEIR TACTICS WORKED. LUMBER COMPANIES HAD TO STOP CUTTING FOR A WHILE.

BUT THAT LED TO EVEN MORE TENSION BETWEEN ACTIVISTS AND LOGGERS, WHO FELT THEIR JOBS WERE BEING THREATENED.

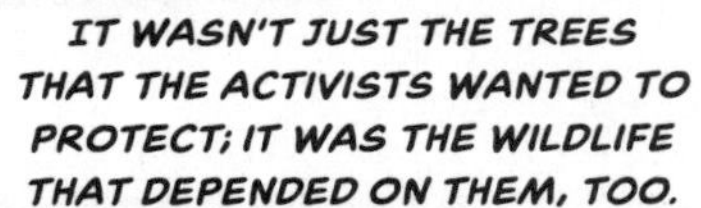

AT FIRST THE GOVERNMENT REFUSED TO PROTECT THE OWLS BUT LISTED THEM AS THREATENED UNDER THE ENDANGERED SPECIES ACT IN 1990. THIS LED TO STRICTER LIMITS ON CUTTING TREES ON NATIONAL FOREST LANDS.

THE DEBATE BECAME A CAMPAIGN ISSUE IN THE 1992 PRESIDENTIAL ELECTION.
IT'S TIME TO MAKE PEOPLE MORE IMPORTANT THAN OWLS.
GOV. GEORGE H.W. BUSH (Texas)
REPUBLICAN PRESIDENTIAL CANDIDATE

WHEN BILL CLINTON WON THAT ELECTION AND BECAME PRESIDENT, HE SUPPORTED A COMPROMISE HE HOPED WOULD SATISFY EVERYONE.
HE PROMISED OVER A BILLION DOLLARS TO RETRAIN LOGGERS AND DEVELOP OTHER INDUSTRIES FOR THE REGION, SO THE OWLS COULD BE SAVED AND THE LOGGERS COULD FIND NEW JOBS.

OWLS!
IT DIDN'T STOP THE ARGUING.
JOBS!

THE ORIGINAL TREE HUGGERS

Tree sitters of the early 1990s weren't the first people to risk their lives to protect a forest. Way back in 1730, India's ruler, the maharaja, ordered some trees in the Himalayan foothills to be chopped down. A woman named Amrita Devi wanted to save the trees, which were sacred to the local people, so she wrapped herself around one of them to protect it.

The king's loggers killed Devi with an axe and did the same to her daughters when they stepped up to take her place. In the end, more than three hundred villagers died trying to protect their trees. When the maharaja heard what was happening, he changed his mind. He called off the whole project and apologized to the people of the village (the ones who were still alive, anyway).

The story of that village inspired India's Chipko movement, a modern forest conservation effort. The Hindi word "chipko" means "to hug," which is why environmentalists are sometimes referred to as "tree huggers" today.

Many environmental battles are fought in specific regions—a landfill in one city, a toxic waste site in another—but in 1985, the whole world faced a challenge when scientists discovered a hole in the ozone layer, the part of Earth's atmosphere that absorbs most of the radiation from the sun. Experts had known for a while that chemicals called chlorofluorocarbons (CFCs) were making the ozone layer thinner. But this new research showed that the damage was happening faster than they thought. A *lot* faster.

At the time, CFCs were common. The chemicals were used in air conditioners, refrigerators, hair spray, furniture polish, and other products. But now scientists understood that those everyday chemicals were causing a serious problem.

A thinning ozone layer meant more of the sun's harmful ultraviolet light could reach Earth. That was dangerous, increasing people's risk of everything from sunburns to skin cancer to cataracts, a condition in which the lens of the eye becomes cloudy.

World leaders acted quickly. In 1987, the United States and twenty-three other nations signed an agreement called the Montreal Protocol—a promise

to begin phasing out the production and use of CFCs. The rapid response was effective, and the ozone layer has begun to recover.

But as that problem was being solved, a greater threat was looming. Soon after the Montreal Protocol was signed, scientists raised the alarm about an even more serious issue—one that has been making headlines ever since.

MOST LIKELY TO PLANT A TREE (AND THEN PLANT FIFTY MILLION MORE)

My greatest satisfaction is to look back and see how far we have come. Something so simple, but meaning so much, something nobody can take away from the people, something that is changing the face of the landscape. . . . I never knew when I was working in my backyard that what I was playing around with would one day become a whole movement. One person can make a difference.

WANGARI MAATHAI was a Kenyan scientist who chaired her country's National Council of Women from 1981 to 1987. In this role, she introduced the importance of planting trees to help the environment and improve people's quality of life. She launched the Green Belt Movement, which has since planted more than fifty million trees in Kenya.

THE ~~HAPPY~~-CAMPER AWARD
UNSTOPPABLE

JOANN TALL became an environmental activist to protect the lands of her people, the Oglala Lakota tribe in the Pine Ridge Reservation of South Dakota. In 1987, a company called Honeywell announced a plan to conduct nuclear weapons tests in the Black Hills, a sacred site for Lakota people.

Tall owned a radio station at the time, and she used it to spread the word about what was happening to her tribe. She led more than a hundred Lakota people to camp out in the canyon for three months to block the project. They refused to leave until Honeywell backed off the testing plan.

MOST LIKELY TO SAVE THE RAINFOREST

CHICO MENDES was a Brazilian labor leader and conservationist who fought against deforestation in the Amazon rainforest. He grew up tapping rubber trees with his dad in the 1950s and 1960s, at a time when many tappers were being forced off their land so it could be cleared and burned to make pastures for cattle.

That wasn't good for the rubber tree tappers or the forest, so Mendes stepped up to organize a labor union. The tree tappers joined forces with others who wanted to save the rainforest. They presented their case to the government and even stood in front of tractors and chain saws in an attempt to stop the clear-cutting.

Mendes fought for extractive forest reserves—territories where products like rubber could be extracted using sustainable methods that protected both the trees and the rights of poor workers and Indigenous people who relied on them.

In 1988, Mendes was murdered by a rancher who opposed his efforts. That same year, Brazil created the first extractive forest reserve.

Nine
OUR WARMING WORLD

In 1988, a top NASA climate scientist alerted US lawmakers that the world's climate was changing. James Hansen, the director of NASA's Goddard Institute for Space Studies, shared compelling research from numerous scientists that confirmed global warming was underway. And it was caused by human activity.

—FROM JAMES HANSEN'S TESTIMONY BEFORE THE US SENATE IN 1988

By the 1980s, scientists also understood *why* the world was getting warmer. Carbon dioxide (CO_2) from the burning of fossil fuels traps the sun's radiation close to Earth, creating a greenhouse effect. CO_2 and other gases act like a blanket, warming the surface of the earth. We call these greenhouse gases.

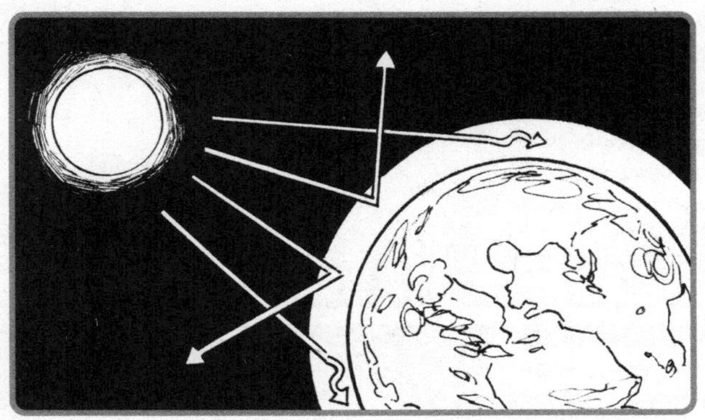

Some of the sun's energy is absorbed by gases surrounding Earth, which can heat our planet. This is called the greenhouse effect.

That information wasn't new when Hansen testified in 1988. Seven years earlier, he and other researchers had warned that the greenhouse effect would lead to a noticeably warmer planet within the decade.

BREAKING NE

AUGUST 22, 1981

Study Finds Warming Trend That Could Raise Sea Levels

—FROM *THE NEW YORK TIMES*

And the greenhouse effect wasn't exactly breaking news in 1981, either. Scientists had actually discovered it more than a hundred years earlier.

THEN SHE PUT A THERMOMETER IN EACH ONE AND LEFT THEM IN THE SUN.

SHE FOUND THAT THE CYLINDER WITH CARBON DIOXIDE WARMED THE MOST . . .
. . . AND TOOK A LONG TIME TO COOL, EVEN AFTER IT WAS MOVED OUT OF THE SUN.

SHE KNEW CARBON DIOXIDE WAS ONE OF THE GASES IN EARTH'S ATMOSPHERE.
AND SHE UNDERSTOOD THAT HER RESEARCH COULD BE IMPORTANT.

An atmosphere of that gas would give to our earth a high temperat

EUNICE FOOTE HAD DISCOVERED GLOBAL WARMING.

BUT IN THOSE DAYS, WOMEN DIDN'T OFTEN PRESENT AT SCIENTIFIC CONFERENCES. SO A MALE COLLEAGUE SHARED EUNICE FOOTE'S RESEARCH AT A BIG MEETING OF SCIENTISTS.
HE DIDN'T UNDERSTAND HOW SIGNIFICANT EUNICE'S FINDINGS WERE.
SO HER WORK GOT LEFT OUT OF THE SUMMARY OF THE MEETING.
SHE EVENTUALLY PUBLISHED HER PAPER, BUT NO ONE PAID MUCH ATTENTION TO THE ISSUE UNTIL JOHN TYNDALL DID A SIMILAR EXPERIMENT A FEW YEARS LATER.

A Swedish chemist named Svante Arrhenius picked up where Tyndall left off. He wondered about the impact of carbon dioxide on the climate. If more CO_2 made things warmer, could falling levels of CO_2 have caused the ice age?

Arrhenius did some calculations about how changes in CO_2 levels might affect Earth's atmosphere. It wasn't perfect research—he did it all with

Swedish chemist Svante Arrhenius, around 1910

pen and paper and made a bunch of mistakes. But his mistakes ended up mostly canceling themselves out, and he came to the right conclusion: if CO_2 levels were to double, then Earth's average temperature would likely rise by nine to eleven degrees. That's pretty close to the estimates we have from modern computer models. Arrhenius also recognized that climate change was linked to industrialization—meaning that all those coal-burning factories, railroads, and power plants were the primary cause.

But here's what Arrhenius got wrong: he thought the oceans would soak up all that extra CO_2, so any gas buildup would happen very, very slowly, after *thousands* of years of burning fossil fuels. In the meantime, he suggested that it might actually be a good thing, since a slight rise in average temperatures would mean a longer growing season. More crops! Yay!

Arrhenius died in 1927, and for a while, scientists weren't terribly concerned about climate change. Why worry about something that wouldn't happen for thousands of years?

But it turns out the climate was changing a lot faster than anyone had predicted. In the 1950s, chemist Charles David Keeling developed a better way to measure CO_2 in the atmosphere.

President George W. Bush awards the National Medal of Science to Charles David Keeling in 2002.

The US Weather Bureau started using Keeling's technique to monitor CO_2 levels at its new observatory on Mauna Loa, which stands at eleven thousand feet above sea level in Hawaii.

The Mauna Loa Observatory

Measurements have been taken there since 1958, and the results show that Arrhenius was way off on his timeline. The data collected reflects a steep rise in atmospheric CO_2 levels, aptly named a Keeling Curve. The graph shows that Arrhenius's predictions about climate change are happening over a much shorter period of time—about one hundred fifty years instead of three thousand.

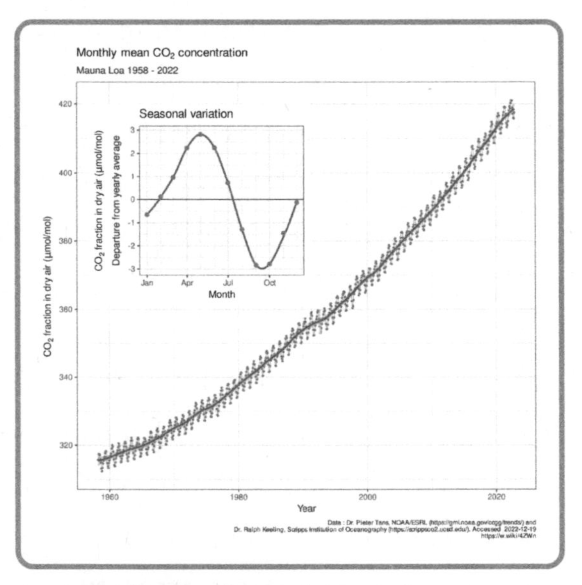

The Keeling Curve shows an increase in CO_2 concentration in the upper atmosphere from 1958 to 2022.

Climate scientists continued to study greenhouse gases throughout the 1960s and 1970s. They created

Akira Kasahara and Warren Washington at the National Center for Atmospheric Research

mathematical models to predict how the earth's climate might change as a result of how much additional CO_2 was in the atmosphere.

Akira Kasahara and Warren Washington started work at the National Center for Atmospheric Research in 1963. Together, they created one of the first general circulation computer models (GCMs) of the atmosphere.

GCMs are models that simulate Earth's climate system. They're used for weather forecasting, understanding climate patterns, and making predictions about climate change. The models that Kasahara and Washington developed paved the way for scientists to study the impact of human activities on the climate.

By 1988, when Hansen and a panel of other scientists testified before Congress, researchers had already learned a lot about global warming—enough to say with certainty that human activities were causing climate change and that it was happening fast enough to be a major concern. It was front-page news.

BREAKING NEW

JUNE 24, 1988

If Dr. Hansen and other scientists are correct, then humans, by burning of fossil fuels and other activities, have altered the global climate in a manner that will affect life on earth for centuries to come.

—FROM *THE NEW YORK TIMES*

Scientists on the panel called for major cuts in the burning of coal, oil, and other fuels that release carbon dioxide. They urged nations around the world to stop clear-cutting trees and plant more of them. (Trees absorb and store CO_2, so it doesn't end up in the atmosphere.)

So what happened next? You might think this big news would prompt big changes in the use of fossil fuels around the world. But the world had been relying on fossil fuels for more than a century, and not everyone was ready to make major changes.

The scientists kept speaking up, though, and some policymakers began to take the issue seriously. In 1988, the United Nations formed an Intergovernmental Panel on Climate Change (IPCC) to advance scientific knowledge about climate change. The group issued reports on its findings, which confirmed over and over again that climate change due to human activities is causing the planet to warm to dangerous levels.

In June 1992, world leaders gathered for an Earth Summit, a first-of-its-kind meeting to address global environmental issues, including climate

change. Sounds promising, right? But the summit ended with no specific plan for cutting global carbon dioxide emissions.

In 1997, the United Nations held another conference about climate change in Kyoto, Japan. Once again, leaders discussed plans to reduce greenhouse gases. US Vice President Al Gore delivered an opening address.

This time, leaders from wealthy countries talked a lot about how developing nations—countries where more citizens live in poverty—should have to help with the efforts to combat climate change. Meanwhile, those developing nations argued that wealthier countries, including the United States, had pumped a lot of greenhouse gases into the atmosphere while they were busy building factories and getting rich. Why should the poorer nations have to help clean up that mess?

The conference ended with an agreement called the Kyoto Protocol, which set binding and specific limits on greenhouse gas emissions for industrialized nations. It took effect in 2005.

—DR. MARK MWANDOSYA, TANZANIA

Finally, some meaningful progress! Right?

Not really. Most nations ended up missing their targets. But an even bigger problem was that the world's two largest greenhouse gas emitters weren't part of the agreement at all. China was exempted because it was considered a developing nation, and the United States refused to join. (The United States technically signed the agreement, but President Clinton never submitted it to the Senate for approval. He said he couldn't support it without participation from developing nations and was also concerned about jobs.)

Another agreement called the Paris Accord replaced the Kyoto Protocol in 2015, once again setting limits on carbon emissions to slow global warming. This time, nearly two hundred nations joined, including the United States.

But support for the agreement has gone back and forth. When President Donald Trump took office in 2017, he withdrew the United States from the accord, saying that the agreement would hurt the economy and workers and put the United States at a disadvantage compared to other nations. The United States rejoined the Paris Accord four years later, when President Joe Biden took office.

By now, you might be starting to notice that the United States does a lot of flip-flopping when it comes to environmental issues. It all depends on which leaders are in office and how much importance they place on the environment compared with other issues, such as jobs and economic growth.

Companies that earn money from fossil fuels and the people who rely on those companies for jobs generally don't support big changes that could affect their business. Those same companies donate a lot of cash to political campaigns, hoping that elected leaders will vote for policies that will help those companies make more money. Meanwhile, environmentalists argue that it's past time to stop relying on fossil fuels like coal and oil, which harm the environment in more ways than one.

ON AGAIN, OFF AGAIN

Speaking of flip-flopping, remember those solar panels on the White House roof? The ones President Carter installed and President Reagan took down? President Barack

Obama ordered new solar panels for the roof after he took office, and the work was completed in 2014.

But it was actually President George W. Bush who installed the first solar power system for the White House grounds. In 2003, he had panels installed on a maintenance building to heat the White House swimming pool.

Support for environmental protection tends to surge after big eco-disasters. On April 20, 2010, British Petroleum's Deepwater Horizon oil-drilling rig exploded, killing eleven people and spilling more than a hundred million gallons of oil into the Gulf of Mexico. It devastated the region's fishing and tourism industries, as well as its wildlife.

Crews fight fires after the Deepwater Horizon explosion.

A veterinarian tries to help an oil-coated sea turtle after the 2010 Deepwater Horizon oil spill.

The oil company had to pay a fine, and for a while, the United States didn't issue new permits for offshore drilling. Demonstrators called for a boycott of British Petroleum, but others argued that oil was essential to the economy, and it was important to keep oil prices low, even if there were disasters once in a while.

WHAT THE FRACK?

Environmentalists and energy companies have also battled over an oil extraction process called fracking. It involves injecting liquid at high pressure into underground

rocks to create cracks for extracting oil or gas. The process creates large amounts of wastewater and also emits greenhouse gases such as methane. But groups that support fracking argue that it allows access to gas and oil deposits that were previously impossible to reach, and that can lead to lower energy prices.

There were also protests over the expansion of the Keystone XL Pipeline, which a company called TC Energy was building to transport oil from tar fields in Canada to oil refineries in Texas. The pipeline's planned route would have taken it through Indigenous lands and an aquifer that provides water for millions of people in the Plains states.

America's leaders went back and forth about the project. President Obama vetoed the pipeline in 2015, saying it was a threat to climate, wildlife, and drinking water. President Trump revived the project in 2017, saying it would create thousands of good-paying construction jobs that Americans needed. President Biden pulled support for the project on his first day in office, denying a permit that was necessary for it to continue. After a decade of controversy, TC Energy suspended work in January 2021 and ultimately canceled the project five months later.

Protesters also fought plans for an oil pipeline near the Standing Rock Sioux Reservation. The tribe argued that the proposed pipeline violated a treaty that

Dakota Pipeline protesters gathered outside Minneapolis City Hall on October 25, 2016.

guaranteed them "undisturbed use and occupation" of reservation lands that surrounded the pipeline's proposed site. Hundreds of people attended marches, horseback rides, runs, and protests of all kinds.

The project was delayed, so the Army Corps of Engineers could perform more research on the environmental impact, but when President Trump took office, he ordered that the review and approval process be sped up. In the end, the pipeline was completed, but the people who fought against it sparked a larger "water protectors" movement, inspiring other groups to fight projects that they believe threaten their water supplies.

And through it all, scientists continued to sound the alarm about fossil fuels and global warming. The impacts were real, and they were growing more alarming every day.

Ten
SOLUTIONS FOR TOMORROW

Since James Hansen and his fellow scientists testified before Congress in 1988, NASA has done hundreds of additional studies on climate change. They all came to the same conclusion: human activities are causing the climate to warm in ways that are already having an impact on the earth and its oceans.

In 1979, NASA did a study of the Arctic's perennial sea ice—the kind that's frozen all year, not just in winter. The ice covered 1.7 billion acres of ocean—about the size of the continental United States. Imagine an area that large covered in ice.

Now imagine chopping off New York, Georgia,

and Texas. That's how much the ice had shrunk by 2023—more than 250 million acres.

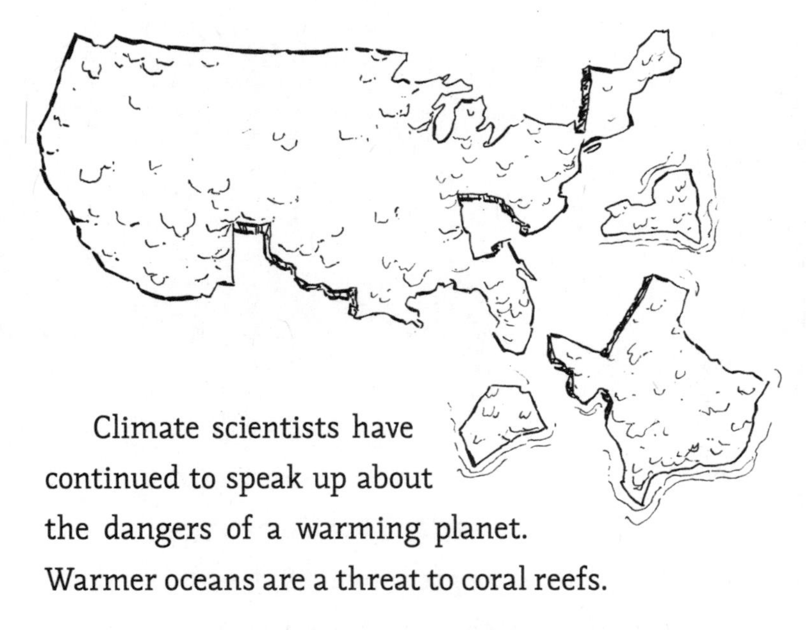

Climate scientists have continued to speak up about the dangers of a warming planet. Warmer oceans are a threat to coral reefs.

Coral reefs have been around for five hundred million years and support one quarter of all marine species. But they're in danger because of climate change and pollution.

Higher temperatures have also melted sea ice, putting polar bears at risk.

Polar bears depend on sea ice; it's where they hunt for seals, their main source of food.

With warmer temperatures, disease-carrying insects can migrate farther north, reaching more people who could be affected by illnesses such as malaria.

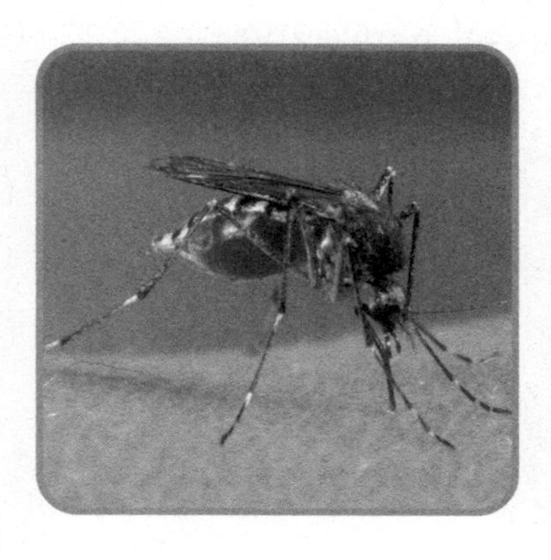

Nine cases of locally transmitted malaria were reported in Florida, Texas, and Maryland in the summer of 2023. It was the first time in twenty years that a case was reported in a United States resident who hadn't been traveling recently.

Warmer weather also causes more dangerous storms. Combined with higher sea levels, that means more frequent and more severe flooding.

Superstorm Sandy hit Cuba on October 25, 2012, then barreled up the east coast of the United States and Canada, killing more than two hundred people and causing almost $70 billion in property damage.

Scientists have also warned that global warming will lead to more frequent heat waves, droughts, and wildfires.

A fire plane battles the August Complex fire on September 25, 2020. Six of the biggest wildfires in US history took place between 2000 and 2023.

Even with all those warnings, some corporations and government officials have done their best to silence scientists and mislead the public. Petroleum companies, for example, knew they'd lose money if people stopped using so much oil. So they spent hundreds of millions of dollars fighting laws that would put limits on carbon emissions. They also paid for their *own* research, hiring scientists who would say what the company wanted to hear about climate change.

It's similar to the fight that big tobacco companies waged when scientists learned how dangerous smoking is to people's health. Those tobacco company "experts" denied that the science was real and instead directed everyone to pay attention to their own hired scientists. Anything to confuse people so they wouldn't worry about smoking.

BIG TOBACCO

Doubt is our product since it is the means of competing with the "body of fact" that exists in the mind of the general public. It is also the means of establishing a controversy.

—FROM A 1969 BROWN AND WILLIAMSON TOBACCO DOCUMENT

In other words: Convincing people to doubt science is the only way tobacco companies can win this one. The more confused people are, the better off business will be.

Exxon Mobil, one of the world's largest oil and gas companies, knew about climate change by the late 1970s. Their own senior scientist told company managers that burning fossil fuels released carbon dioxide, which contributed to global warming. But for decades, Exxon Mobil donated millions of dollars to groups that denied climate science.

Some people are *still* trying to confuse the public with fake science. Here are a few of the climate change myths that need smashing.

The Science: Weather and climate aren't the same thing. If it's very hot or very cold one day, or even for a whole week, that's just weather, and it's not evidence of anything. Climate refers to *long-term* average temperatures based on decades of data. That's the data scientists use to determine how climate change is occurring and what's likely to happen in the future.

THE MYTH: Earth's climate changes all the time, and it's been happening for millions of years. The climate variations are probably natural.

The Science: Earth's climate *has* changed over millions of years, for lots of natural reasons. These include changes in the sun, moving continents, volcanic eruptions, and variations in Earth's orbit around the sun.

But *none* of those things can account for the rapid rise in Earth's average temperature over the past sixty years. That change is a result of human activity.

THE MYTH: Climate change isn't caused by people; it's caused by this thing called the Milankovitch effect. It has something to do with a tilt in Earth's orbit.

The Science: It's true that in 1941, a Serbian mathematician and scientist named Milutin Milankovitch discovered that natural variations in Earth's orbit and tilt have played a role in the development of glaciers over the past two million years.

We know about Milankovitch cycles because they've been recorded in ice cores, samples of glacier ice that contain information about temperatures and atmospheres of the past. Thanks to the work of scientists who study those ice cores, we now understand Earth's orbital cycles pretty well. And they are *not* responsible for the rapid increase in global warming since the 1950s. In fact, during that time, Earth has actually been in a part of the cycle that contributes to *cooling*—not warming.

THE MYTH: El Niño is causing climate change.

The Science: **El Niño is a phenomenon that leads to warmer ocean currents along the western coast of the Americas in certain years. For example, 1998 was a strong El Niño year, so temperatures were warmer than usual. When the next two years were cooler, climate deniers claimed it was evidence that global warming wasn't really happening. By 2005, temperatures were running high again, and those arguments quieted down.**

The truth is, while El Niño can affect temperatures for a few years, there's no evidence that it affects climate in the long term, and it is not the cause of global warming.

At this point, there's no debate among legitimate scientists. Nearly 100 percent of them understand that climate change is real, impacted by humans, and a threat to both wildlife and people. Most Americans now believe in climate change, too. They just don't agree on what should be done about it.

But the hundreds of international climate experts who are part of the IPCC have made their suggestions clear. According to IPCC reports, industrialized nations would need to stop adding greenhouse gases to the atmosphere by 2050 to limit global warming to 1.5 degrees Celsius—enough to temper some of the most serious effects.

Not meeting that goal doesn't mean humans would be doomed. But it does mean that many would face more dangerous situations—from storms, floods, and food shortages to diseases, wildfires, and heat waves—in the years to come.

SWEATING IT OUT OVER CLIMATE CONCERNS

It's possible that reading these last two chapters about climate change has stressed you out a little. (You're not alone. Writing them stressed me out, too.) Learning about all the challenges the world faces because of global warming can feel overwhelming. Especially when you don't have much control over how to fix it.

This feeling is so common that it has a name: climate anxiety. An American Psychological Association survey found that most Americans are worried about the climate, and more than half of children and young adults reported feeling *very* or *extremely* worried.

That concern is understandable. Climate change isn't a good thing. But neither is anxiety, so experts recommend taking steps to manage it. Sometimes that might mean taking a break from reading about climate change or scrolling through the news.

Going outside, getting some exercise, and talking to friends and family can help. So can taking small actions

to fight climate change—whether that involves making posters, going to meetings, or biking to school.

Finally, when you feel anxious or discouraged, try to remember that scientists and thoughtful leaders all over the world are studying this problem—and working on some pretty amazing ideas to help.

We've talked a lot about the bad news surrounding climate change. The good news is that there have also been many positive developments since scientists first raised the alarm. Eighteen nations, including the United States, have been lowering carbon emissions for more than a decade.

The term "net zero" is used to describe a balance between how much carbon is added to the atmosphere and how much is removed. A nation might reach that net zero goal by both reducing its carbon output and engaging in practices that remove CO_2 from the atmosphere. These include low-tech strategies like replanting forests as well as high-tech approaches, like direct air capture plants that use chemical reactions to remove carbon dioxide from the air.

As of 2023, the United States and European Union were both on track to reach net zero by 2050. China was expected to hit that goal by 2060, and India by 2070.

Nations have also made investments in clean, renewable energy. Solar panels, wind turbines, and lithium-ion batteries used in electric vehicles are less expensive than they used to be. More people are driving cars that don't run on gasoline. In 2022, Congress passed a law that included the largest climate investment in history—more than $300 billion to reduce carbon emissions, including incentives to help people transition to green energy.

People are changing their individual habits, too. Plant-based diets are more popular than ever, and they result in fewer greenhouse gas emissions than meat-based diets.

You might not think much about it when you're at the grocery store, but different foods have different impacts on the environment. Meat is actually responsible for a good chunk of the world's greenhouse gas emissions—up to 22 percent. And beef is the number one culprit.

When cows burp and fart, they produce methane. When cows poop, their manure often ends up in settling ponds or lagoons, where solid waste is filtered

out from liquid. These smelly little lakes emit even more methane.

Farming pigs, chicken, and fish also leads to greenhouse gas emissions, though not as much as cows. Vegetables and fruits have less of an impact, which is why many environmentalists choose to eat plant-based diets.

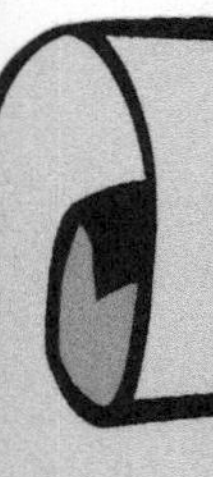

WHAT CAN REGULAR PEOPLE DO ABOUT CLIMATE CHANGE?

Climate change is a big problem that can't be solved without lots of nations coming together and governments committing to cut carbon emissions. But that doesn't mean there's nothing regular people can do to help.

When you're a kid, a lot of choices aren't up to you. Older family members probably decide what's for dinner, where you travel, and how you get there most of the time.

But if the people in charge of the kitchen at your house take requests, you might ask for a meatless meal once in a while. And if your family is making a short trip, you can always suggest walking, biking, or using public transit.

There are lots of ways to cut energy use beyond obvious actions like turning off lights and the TV when you leave a room. If your clothes aren't dirty, you can wear them more than once before throwing them into the wash. And only run the dishwasher when it's full.

You can also try to cut back on the garbage you produce. When waste breaks down in landfills, it emits CO_2 and methane—both greenhouse gases. Recycling, reusing items, composting, and reducing food waste all help to keep trash out of landfills. That's especially important when it comes to things made out of plastic, which are produced using fossil fuels and often end up in the world's oceans.

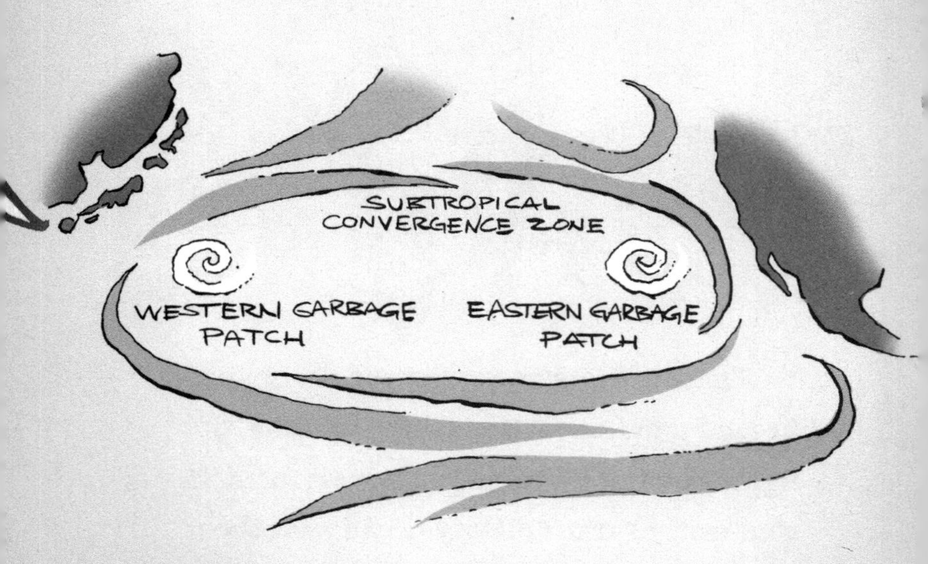

Ocean currents have carried tons of discarded plastics to an area of the North Pacific that's so clogged with debris it's now known as the Great Pacific Garbage Patch. It's twice the size of Texas—and growing. Similar plastics patches have accumulated in other oceans as well.

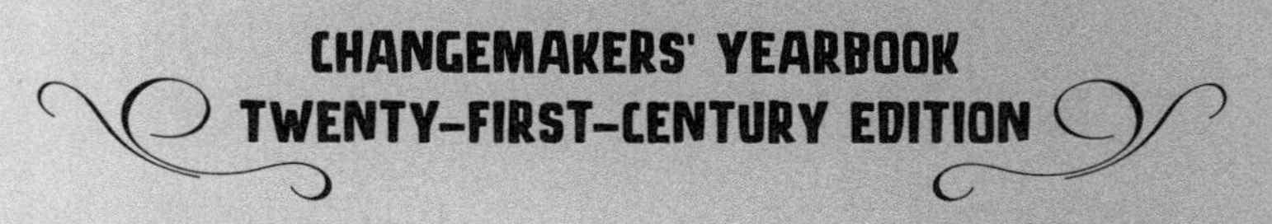

MOST LIKELY TO GET TO THE POINT

Korean economist **HOESUNG LEE** became chair of the Intergovernmental Panel on Climate Change in 2015. His leadership emphasized sustainable development—economic growth that doesn't pump more CO_2 into the atmosphere. He also worked to make the IPCC's lengthy reports easier for non-scientists to understand.

THE LITTLE MISS FLINT AWARD

When **MARI COPENY** was eight years old, she wrote **PRESIDENT OBAMA** a letter about the contaminated drinking water in Flint, Michigan. The water where she lived was making people sick, and studies showed that it contained dangerous levels of bacteria and lead. The president wrote back to Mari and promised to visit Flint himself. Seven months later, Congress passed the Water Infrastructure Improvements for the Nation Act, which included $100 million dedicated to repairing Flint's broken water system. Mari, who became known as Little Miss Flint, went on to become an activist for other causes as well.

THE TAKE-YOUR-STATE-TO-COURT AWARD

Rikki Held, Badge and Lander Busse, Sariel Sandoval, Kian Tanner, Georgianna Fisher, Grace Gibson-Snyder, Eva Lighthiser, Mica Kantor, Olivia Vesovich, Nathaniel and Jeffrey King, Claire Vlases, Lilian and Ruby D., Taleah Hernandez

In 2020, more than a dozen young people went to court to fight climate change. They sued the state of Montana, claiming the state violated their constitutional right to a healthy environment by promoting fossil fuels that contribute to global warming. The court case wrapped up in 2023 with the judge ruling that the kids were right.

"Montana's emissions and climate change have been proven to be a substantial factor in causing climate impact to Montana's environment," and these emissions directly impact the state's youth.

Despite this win, changes didn't happen right away. The court sent the issue to the state legislature to decide how to put new policies in place.

MOST LIKELY TO LIFT UNHEARD VOICES

VANESSA NAKATE is a Ugandan climate activist who led a campaign to save Congo's rainforest. She was chosen to speak at the 2019 COP 25 UN Climate Change Conference in Spain. Nakate also founded the Rise Up Movement, which empowers African voices and brings attention to the fact that developing nations, which have contributed the least to climate change, often face the most dire effects from it.

MOST LIKELY TO SKIP SCHOOL FOR THE EARTH

Swedish climate activist **GRETA THUNBERG** was fifteen years old when she decided to take action on climate change. She staged a protest and, instead of going to school, sat outside the parliament building with a sign that said SCHOOL STRIKE FOR CLIMATE, calling for stronger action on climate change. Thunberg inspired tens of thousands of students to stage their own climate strikes around the world.

THE GREAT-GLACIERS AWARD

ISABELLE VELICOGNA is a glaciologist. Her research on glaciers in Greenland and

Antarctica has made the world aware of significant losses due to global warming. She uses satellite data to track both polar ice sheets and rises in sea levels.

MOST LIKELY TO WARN YOU ABOUT A DROUGHT

Ugandan researcher **CATHERINE LILIAN NAKALEMBE** studies the impact of climate change on the world's food supply. She uses satellite imagery to map agricultural areas and has already seen the effects of droughts and floods on communities in Africa. Her work has led to early-warning systems for droughts so that small farmers have time to prepare.

THE SUPER-CORALS AWARD

EMMA CAMP is an Australian researcher who studies climate change's impact on coral reefs. Her team is researching "super corals" that are able to tolerate extreme temperature changes. Camp hopes to learn more about how to save reefs threatened by warming oceans.

It's true that a certain amount of global warming is going to continue, based on the CO_2 that's already in the atmosphere. The decade after this book is published is likely to be hotter and stormier than the one before. Scientists are working to solve that problem, too. They're developing earlier-warning systems to predict storms, and engineers are designing coastal barriers to protect cities from flooding.

While some animals will likely go extinct due to climate change, some have already begun adapting, literally moving to climates where they can survive. Others have been forced to change what time of year they migrate or hibernate.

Researcher Colin Donihue used a leaf blower to simulate storm winds in an experiment with Caribbean anole lizards.

Some animals' bodies are even evolving to help them survive. Scientists have observed Australian parrots with larger bills, which are used to regulate body temperatures. And lizards in the Caribbean seem to be evolving to deal with the strong winds that come with more frequent hurricanes. Scientists have observed longer front legs and bigger toe pads, which are good for hanging on in a storm.

In the years to come, every living thing will need to adapt to Earth's changing climate. Scientists know that global temperatures will continue to rise for a time, but they can't say for certain how much or what the effects will be. The answer to that question depends on what happens next.

All over the world, scientists and leaders, activists and changemakers are at work right now. They're tracking data, raising awareness, searching for answers, and teaching people about tools and strategies that already exist. With advances in science and technology, researchers are finding innovative ways to address the climate crisis all the time. There may be solutions that no one has thought of yet, which is why everyone's ideas are important.

Including yours.

Because each discovery—each tiny piece of knowledge—has the potential to lead to a brighter future and a better understanding of our irreplaceable world.

A TIMELINE OF ENVIRONMENTAL HISTORY

In the many thousands of years since humans learned to control fire, they have used it to clear land and burn wood for heat, light, cooking, and other purposes. While these actions most certainly resulted in changes to the land, this timeline focuses on the period beginning with colonization and industrialization, when those changes kicked into high gear.

1492—Christopher Columbus explores islands in the Caribbean, beginning the process of conquest and colonization that leads to the fall of the Aztec and Inca empires and the exchange of plants and animals between Europe and the Americas.

1623—Plymouth Colony passes the first known fishery law to protect the herring run.

1626—After forests begin to diminish, Plymouth Colony passes a law to regulate the cutting of timber.

1668—Pilgrims takes steps to manage water pollution in Plymouth Harbor.

1682—William Penn arrives in Pennsylvania and requires colonists receiving land grants to leave one acre of forested land for every five acres they clear.

1769—James Watt is awarded a patent for a more efficient steam engine that helps jump-start Britain's Industrial Revolution.

1825—Work is completed on the Erie Canal, spanning more than three hundred miles, from Albany to Buffalo, New York.

1844—Poet and newspaper editor William Cullen Bryant suggests setting aside a big piece of land in New York City to create "a great municipal park."

1848—Gold is discovered at Sutter's Mill, launching the California gold rush.

1849—The US Department of the Interior is established and given the authority to administer public lands.

1854—Henry David Thoreau's most famous book, *Walden; or, Life in the Woods,* is published.

1856—Eunice Foote publishes a paper about her experiments, which demonstrated that carbon dioxide (CO_2) absorbs heat and concluded that more CO_2 in the atmosphere would change Earth's climate.

1858—Frederick Law Olmsted and Calvert Vaux win a design contest and begin work on New York City's Central Park.

1859—John Tyndall builds a ratio spectrophotometer and tests the absorptive properties of different gases, coming to similar conclusions as Eunice Foote.

1864—President Abraham Lincoln begins the process of preserving Yosemite Valley.

1872—Yellowstone National Park is established. It's America's first national park.

The United States designates April 10 as Arbor Day, a holiday for planting trees.

1878—Massachusetts passes a law to keep untreated industrial waste and sewage from being dumped into rivers and streams.

1882—Florida begins to drain the Everglades to create more farmland.

1885—New York State creates the Niagara Reservation to protect Niagara Falls and sets aside the Adirondack Forest Preserve, which will later become the Adirondack State Park.

1889—Jane Addams and Ellen Gates Starr found Hull-House in Chicago.

1890—Yosemite National Park is established.

1892—John Muir organizes the Sierra Club.

1895—Swedish scientist Svante Arrhenius presents his research on the effects of higher CO_2 levels in the atmosphere on Earth's climate. He estimates that global warming will take thousands of years. (He's wrong.)

1896—Harriet Hemenway rallies the women of Boston to stop wearing feathers in hats and take action to protect birds, leading to the creation of the Massachusetts Audubon Society.

1897—Massachusetts passes a law that bans the trade of wild bird feathers.

1900—Congress passes the Lacey Act, which makes it easier to punish poachers hunting on public lands and bans the shipment of animals that are killed in violation of local laws.

1905—The National Audubon Society is formed.

1908—Grand Canyon National Monument is established.

President Theodore Roosevelt holds a Conference of Governors on the conservation of natural resources.

1913—Congress passes the Weeks-McLean Migratory Bird Act.

Chicago creates a Commission on Waste to address the garbage and waste in the city streets as a result of the meatpacking industry.

Congress approves a plan to dam the Tuolumne River to create a reservoir in Yosemite's Hetch Hetchy Valley. The vote came after years of controversy and debate that raised awareness of environmental issues.

1914—The last passenger pigeon dies, and the species becomes extinct.

World War I begins.

1916—The US National Park Service is established.

1924—The Gila Wilderness in New Mexico is established as the first designated wilderness area in the United States.

1928—Chlorofluorocarbons (CFCs) are invented for use in refrigeration and air-conditioning.

1935—The Black Sunday dust storm—one of many such storms in the 1930s—causes great economic and agricultural damage in the Plains states, prompting Congress to pass the Soil Conservation Act, aimed at protecting farms and preventing future dust storms.

1935—The Wilderness Society is founded.

1936—The National Wildlife Federation is formed.

1939—World War II begins. The military uses the pesticide DDT to control disease-causing mosquitoes as battles are fought around the world.

1940—The US Fish and Wildlife Service is created.

1947—Marjory Stoneman Douglas publishes *The Everglades: River of Grass*. Everglades National Park is established in Florida later that year.

1948—Heavy smog in Donora, Pennsylvania, kills twenty people and sickens thousands more.

Congress passes the Federal Water Pollution Control Act, which is amended in 1972 and becomes known as the Clean Water Act.

1955—Congress passes the Air Pollution Control Act.

1958—Equipment to measure CO_2 levels is installed on Mauna Loa Observatory in Hawaii.

1962—Rachel Carson's book *Silent Spring* is published, sparking protests against the use of pesticides such as DDT.

President John F. Kennedy holds a White House Conference on Conservation.

1963—The Clean Air Act of 1963 is signed into law.

Akira Kasahara and Warren Washington begin work at the National Center for Atmospheric Research, where they collaborate to build one of the first general circulation computer models of the atmosphere.

1964—The Wilderness Act designates more than eight hundred protected wilderness areas in the United States.

1966—Cesar Chavez and Dolores Huerta bring Mexican American and Filipino American farmworkers together to form the United Farm Workers. One of the organization's goals is to protect field workers from dangerous pesticides.

1967—The Environmental Defense Fund is established to fight environmental battles in court.

1969—A massive oil spill near Santa Barbara, California, blackens beaches and kills wildlife.

The polluted Cuyahoga River catches fire in Ohio.

1970—The first Earth Day is held on April 22.

The National Oceanographic and Atmospheric Administration, the Environmental Protection Agency (EPA), and the Natural Resources Defense Council are established.

President Richard Nixon signs the Clean Air Act of 1970 into law.

Dutch scientist Paul Crutzen warns that human actions may damage Earth's ozone layer.

1971—Greenpeace is founded and begins to take more radical actions on environmental issues.

1972—The United States bans most uses of DDT.

The United Nations Conference on the Human Environment is held in Stockholm, Sweden.

Congress revises the Federal Water Pollution Control Act of 1948 and gives the EPA authority to set standards for wastewater. The revised law is known as the Clean Water Act.

The Marine Mammal Protection Act is enacted.

1973—Congress passes the Endangered Species Act.

An oil embargo by the Organization of Petroleum Exporting Countries sends gasoline and heating oil prices soaring in the United States.

1974—Congress passes the Safe Drinking Water Act.

1975—The Eastern Wilderness Areas Act is signed into law, designating new wilderness areas in the eastern half of the United States.

1976—The Federal Land Policy and Management Act gives the Bureau of Land Management more powers to protect federal lands.

The Toxic Substances Control Act puts the EPA in charge of protecting people and the environment from toxic chemicals and requires manufacturers to run safety tests.

1977—The US Department of Energy is formed.

1978—Hundreds of people are forced to evacuate their homes in the Western New York neighborhood of Love Canal because of toxic waste.

1979—The National Academy of Sciences issues a decisive report about the impact of CO_2 in the atmosphere, concluding that if levels continued to rise, significant climate change would result.

1980—The Alaska National Interest Lands Conservation Act protects more than a hundred million acres of Alaska wilderness.

The Superfund is created to respond to environmental emergencies and manage cleanup of the worst hazardous waste sites in the United States.

1981—NASA scientists publish a paper in the journal *Science,* warning that CO_2 in the atmosphere will lead to a noticeably warmer planet within the decade.

1985—Scientists discover a hole in the ozone layer, and a United Nations conference on ozone depletion is held in Vienna. The Vienna Convention for the Protection of the Ozone Layer is adopted to "protect human health and the environment against adverse effects resulting from modifications of the ozone layer."

1987—The United States and twenty-three other nations sign the Montreal Protocol on Substances that Deplete the Ozone Layer and commit to phasing out CFC production and use.

Joann Tall and other Lakota people protest Honeywell's plans to conduct nuclear weapons tests in the Black Hills. Honeywell cancels the tests.

1988—Top NASA scientists testify to Congress, warning that the world's climate is changing as a result of human activities, and share research showing that the process is already underway.

1989—The *Exxon Valdez* oil tanker runs aground and spills eleven million gallons of oil into Alaska's Prince William Sound.

1990—The Oil Pollution Act requires oil companies to use double-hull tankers to prevent spills and strengthens the EPA's ability to respond when spills happen.

Activists protest the logging of old-growth forests in a movement called Redwood Summer.

1992—The UN Conference on Environment and Development is held in Rio de Janeiro, Brazil.

President George H. W. Bush signs the Elwha River Ecosystem and Fisheries Restoration Act, ordering a dam removal to help salmon populations recover.

1997—A UN conference on global climate change is held in Kyoto, Japan, resulting in the Kyoto Protocol, an agreement among many industrialized nations to reduce greenhouse gas emissions.

2002—The Larsen B ice shelf collapses in Antarctica due to a series of warm summers.

2003—The American Geophysical Union issues a consensus statement that "natural influences cannot explain the rapid increase in global near-surface temperatures."

2005—The Kyoto Protocol goes into effect.

2006—Researchers report that loss of ice from Greenland has doubled in the last ten years.

2007—The US Supreme Court rules that the EPA has the power to regulate CO_2 emissions based on the Clean Air Act.

The Intergovernmental Panel on Climate Change (IPCC) issues its fourth assessment report, stating that "warming of the climate is unequivocal" and mostly due to human activity.

2010—British Petroleum's Deepwater Horizon oil drilling rig explodes, spilling more than a hundred million gallons of oil into the Gulf of Mexico.

2012—Superstorm Sandy hits the East Coast, resulting in over two hundred deaths and billions of dollars in property damage.

2014—NASA scientists report that the West Antarctic ice sheet has started to melt and that the process will be irreversible.

The IPCC's fifth assessment report warns that many species "will not be able to move fast enough during the twenty-first century to track suitable climates."

The EPA proposes limiting CO_2 emissions from power plants.

2015—Nearly two hundred nations, including the United States, join the Paris Accord, committing to limit carbon dioxide emissions in an effort to slow global warming.

President Barack Obama vetoes the Keystone XL Pipeline.

2016—Members of the Standing Rock Sioux tribe and others protest the Dakota Access Pipeline, which is eventually put on hold pending a review from the Army Corps of Engineers.

2017—President Donald Trump takes office. He withdraws the United States from the Paris Accord, approves the Keystone XL Pipeline, and orders the Dakota Access Pipeline plan to be reviewed quickly and approved.

2018—Fifteen-year-old Greta Thunberg launches a youth climate strike by skipping school to protest outside Sweden's parliament building, calling for more action on climate change.

2020—California experiences its worst wildfire year in modern history.

More than a dozen young people file a lawsuit against the state of Montana for violating their constitutional right to a healthy environment by promoting fossil fuels that lead to global warming.

2021—President Joe Biden takes office, and the United States rejoins the Paris Accord. Biden also refuses a permit needed for the Keystone XL Pipeline to continue, and the project is canceled.

2022—Congress passes the Inflation Reduction Act, which includes the largest climate investment in US History—$370 billion to help reduce carbon emissions and transition to green energy.

2023—Nine cases of locally transmitted malaria are reported in the United States.

Smoke from wildfires in Canada blankets the northeastern and midwestern parts of the United States, leading to unhealthy air quality warnings for millions.

The World Meteorological Organization declares 2023 the hottest year on record and warns of increasing floods, wildfires, heat waves, and melting glaciers in the future.

AUTHOR'S NOTE

The sky outside my home office in upstate New York is a weird, hazy yellow-orange-gray as I draft this author's note. It looks a little like Mars out there.

It's June 2023, and the air smells of smoke drifting down from wildfires that have been burning for weeks in eastern Canada. Millions of people in the United States woke up to air quality alerts this morning—warnings that it wouldn't be healthy to spend much time outside.

Days like this one are likely to be more common in the years ahead. Increases in global average temperatures have already led to more droughts and wildfires, as well as a host of other challenges, many of which you read about in this book. It can be scary to read about all the ways our planet has already been hurt, and all the warning signs going off right now as the earth still needs our help. And it can be discouraging when it feels like change isn't happening fast enough.

But I hope that you'll also find inspiration in the pages of this book, in the stories of people who have spoken up for the environment over hundreds of years. Humans have tackled environmental problems before. Just as some animals are already adapting to climate change, humans can also adapt. People are capable of changing laws, changing their habits, and changing their beliefs.

It's true that the small things we do can't fix the problem. Global warming isn't going to be solved by a handful of us biking to school or turning off the water while we brush our teeth. But small acts, especially by young people, often inspire the adults around them to do better. And when enough ordinary people decide that change is needed, government leaders can't help but pay attention.

When I set out to write this book in the History Smashers series, I knew there would be no way to cover centuries of environmental history in less than two hundred pages. While this book offers an overview of some important events and the people who have shaped them, it also leaves a lot out. Here are some resources to check out if you'd like to keep exploring these ideas and learn about other environmental issues and changemakers.

BOOKS ABOUT CLIMATE CHANGE, CLIMATE SCIENTISTS, CLIMATE ACTIVISTS, AND WHAT KIDS CAN DO TO HELP

Breaking the Mold: Changing the Face of Climate Science by Dana Alison Levy (Holiday House, 2023)

Climate Warriors: Fourteen Scientists and Fourteen Ways We Can Save Our Planet by Laura Gehl (Millbrook, 2023)

Drawn to Change the World Graphic Novel Collection: 16 Youth Climate Activists, 16 Artists by Emma Reynolds and illustrators (HarperAlley, 2023)

Earth Heroes: Twenty Inspiring Stories of People Saving Our World by Lily Dyu, illustrated by Jackie Lay (Nosy Crow, 2019)

Nature's Best Hope (Young Readers Edition): How You Can Save the World in Your Own Yard by Douglas W. Tallamy, adapted by Sarah L. Thomson (Timber Press, 2023)

No World Too Big: Young People Fighting Global Climate Change by Lindsay H. Metcalf, Keila V. Dawson, and Jeanette Bradley (Charlesbridge, 2023)

One Earth: People of Color Protecting Our Planet by Anuradha Rao (Orca Books, 2020)

Saving Earth: Climate Change and the Fight for Our Future by Olugbemisola Rhuday-Perkovich, introduction by Nathaniel Rich, illustrated by Tim Foley (Farrar, Straus and Giroux, 2022)

BOOKS ABOUT SAVING WILDLIFE

The Brilliant Deep: Rebuilding the World's Coral Reefs by Kate Messner, illustrated by Matthew Forsythe (Chronicle Books, 2018)

Crossings: Extraordinary Structures for Extraordinary Animals by Katy S. Duffield, illustrated by Mike Orodán (Beach Lane Books, 2020)

The Late, Great Endlings: Stories of the Last Survivors by Deborah Kerbel, illustrated by Aimée van Drimmelen (Orca Books, 2022)

Pika Country: Climate Change at the Top of the World by Dorothy Hinshaw Patent and Marlo Garnsworthy, photographs by Dan Hartman (Web of Life Children's Books, 2020)

Rewilding: Bringing Wildlife Back Where It Belongs by David A. Steen, illustrated by Chiara Fedele (Neon Squid, 2022)

A River's Gifts: The Mighty Elwha River Reborn by Patricia Newman, illustrated by Natasha Donovan (Millbrook, 2023)

Sea Otter Heroes: The Predators That Saved an Ecosystem by Patricia Newman (Millbrook, 2017)

Superpod: Saving the Endangered Orcas of the Pacific Northwest by Nora Nickum (Chicago Review Press, 2023)

Tracking Tortoises: The Mission to Save a Galapagos Giant by Kate Messner, photographs by Jake Messner (Millbrook, 2022)

The Wolves and Moose of Isle Royale: Restoring an Island Ecosystem by Nancy Castaldo (Clarion Books, 2022)

World Without Fish by Mark Kurlansky, illustrated by Frank Stockton (Workman, 2011)

BOOKS ABOUT GARBAGE AND WASTE

The Great Stink: How Joseph Bazalgette Solved London's Poop Pollution Problem by Colleen Paeff, illustrated by Nancy Carpenter (Margaret K. McElderry Books, 2021)

Is It Okay to Pee in the Ocean?: The Fascinating Science of Our Waste and Our World by Ella Schwartz, illustrated by Lily Williams (Bloomsbury, 2023)

One Plastic Bag: Isatou Ceesay and the Recycling Women of the Gambia by Miranda Paul, illustrated by Elizabeth Zunon (Millbrook, 2015)

Total Garbage: A Messy Dive into Trash, Waste, and Our World by Rebecca Donnelly, illustrated by John Hendrix (Henry Holt, 2023)

Tracking Trash: Flotsam, Jetsam, and the Science of Ocean Motion by Loree Griffin Burns (Houghton Mifflin, 2007)

PICTURE BOOKS

These all-ages books are great to share with younger friends and family members, too!

Dear Earth . . . From Your Friends in Room 5 by Erin Dealey, illustrated by Luisa Uribe (HarperCollins, 2020)

Mario and the Hole in the Sky: How a Chemist Saved Our Planet by Elizabeth Rusch, illustrated by Teresa Martinez (Charlesbridge, 2019)

One World: 24 Hours on Planet Earth by Nicola Davies, illustrated by Jenni Desmond (Candlewick, 2023)

To Change a Planet by Christina Soontornvat, illustrated by Rahele Jomepour Bell (Scholastic, 2022)

We Are Water Protectors by Carole Lindstrom, illustrated by Michaela Goade (Roaring Brook Press, 2020)

WEBSITES

bioGraphic, from the California Academy of Sciences
biographic.com

Climate Change, from National Geographic Kids
kids.nationalgeographic.com/science/article/climate-change

Climate Change Resources, from WWF
wwf.org.uk/get-involved/schools/resources/climate-change-resources#resources

ClimateKids, from the Earth Science Communications Team at NASA's Jet Propulsion Laboratory
climatekids.nasa.gov

Kids Against Climate Change, from NOAA's Planet Stewards Education Project
kidsagainstclimatechange.co/start-learning

SELECTED BIBLIOGRAPHY

Ashworth, James. "Animals 'Shapeshifting' to Adapt to Rising Temperatures." Natural History Museum. September 16, 2021. nhm.ac.uk/discover/news/2021/september/animals-shapeshifting-to-adapt-to-rising-temperatures.html.

Attfield, Robin. *Environmental Thought: A Short History*. Cambridge: Polity Press, 2021.

Breton, Mary Joy. *Women Pioneers for the Environment*. Boston: Northeastern University Press, 1998.

Bryant, William Cullen. "A New Public Park." *Evening Post* (New York), July 3, 1844.

Buis, Alan. "When Climate Gets Under Your Skin." NASA. May 15, 2023. climate.nasa.gov/news/3267/when-climate-gets-under-your-skin.

Carson, Rachel, *Silent Spring*. Boston: Houghton Mifflin, 1962.

Colbert, Angela. "A Global Biodiversity Crisis: How NASA Satellites Help Track Changes to Life on Earth." NASA. May 22, 2023. climate.nasa.gov/news/3265/a-global-biodiversity-crisis-how-nasa-satellites-help-track-changes-to-life-on-earth.

Crane, Jeff. *The Environment in American History: Nature and the Formation of the United States*. New York: Routledge, 2015.

Donihue, Colin M., Anthony Herrel, Anne-Claire Fabre, Ambika Kamath, Anthony J. Geneva, Thomas W. Schoener, Jason J. Kolbe, and Jonathan B. Losos. "Hurricane-Induced Selection on the Morphology of an Island Lizard." *Nature* 560, no. 7716 (August 2, 2018): 88–91. doi.org/10.1038/s41586-018-0352-3.

Douglas, Marjory Stoneman. *The Everglades: River of Grass*. New York: Rinehart, 1947.

Earle, Steven. *A Brief History of the Earth's Climate: Everyone's Guide to the Science of Climate Change*. Gabriola Island, BC: New Society Publishers, 2021.

Erdman, Shelby Lin. "Study: Global Warming Sparked by Ancient Farming Methods." CNN. August 18, 2009. cnn.com/2009/TECH/science/08/18/ancient.global.warming/index.html.

Foote, Eunice. "Circumstances Affecting the Heat of the Sun's Rays." *American Journal of Science and Arts* 22, no. 66 (November 1856): 382–383. ia800802.us.archive.org/4/items/mobot31753002152491/mobot31753002152491.pdf.

Giggs, Rebecca. "Animals of the Future." *Atlantic*, November 9, 2021. theatlantic.com/magazine/archive/2021/12/animals-adapting-climate-change/620532.

Hansen, James. *Storms of My Grandchildren: The Truth About the Coming Climate Catastrophe and Our Last Chance to Save Humanity*. New York: Bloomsbury, 2009.

Hanson, Thor. *Hurricane Lizards and Plastic Squid*. New York: Basic Books, 2021.

Huddleston, Amara. "Happy 200th Birthday to Eunice Foote, Hidden Climate Science Pioneer." NOAA. July 19, 2019. climate.gov/news-features/features/happy-200th-birthday-eunice-foote-hidden-climate-science-pioneer.

Jackson, Roland. "Eunice Foote, John Tyndall, and a Question of Priority." *Notes and Records: Royal Society Journal of the History of Science* 74, no.1 (March 20, 2020): 105–118. doi.org/10.1098/rsnr.2018.0066.

Jacobo, Julia. "Meet Some of the Women Who Are Fighting Against Climate Change." ABC News. April 22, 2021. abcnews.go.com/US/meet-women-fighting-climate-change/story?id=77030351.

Kline, Benjamin. *First Along the River: A Brief History of the U.S. Environmental Movement*. 4th ed. Lanham, MD: Rowman & Littlefield, 2011.

Kolbert, Elizabeth. *Field Notes from a Catastrophe*. Rev. ed. New York: Bloomsbury, 2015.

Kolbert, Elizabeth. *The Sixth Extinction: An Unnatural History.* New York: Henry Holt, 2014.

Learn, Joshua Rapp. "How the Ancient Maya Practiced Sustainable Agriculture." *Discover.* July 29, 2022. discovermagazine.com/the-sciences/how-the-ancient-maya-practiced-sustainable-agriculture.

Lee, Hoesung, and José Romero, eds. *Climate Change 2023: Synthesis Report. Contribution of Working Groups I, II and III to the Sixth Assessment Report of the Intergovernmental Panel on Climate Change.* doi.org/10.59327/IPCC/AR6-9789291691647.

Lindsey, Rebecca, and Luann Dahlman. "Climate Change: Global Temperature." NOAA. January 18, 2024. climate.gov/news-features/understanding-climate/climate-change-global-temperature.

Montrie, Chad. *The Myth of "Silent Spring": Rethinking the Origins of American Environmentalism.* Oakland: University of California Press, 2018.

Nakate, Vanessa. *A Bigger Picture: My Fight to Bring a New African Voice to the Climate Crisis.* New York: Mariner Books, 2021.

NASA. "2022 Arctic Summer Sea Ice Tied for 10th-Lowest on Record." September 22, 2022. climate.nasa.gov/news/3213/2022-arctic-summer-sea-ice-tied-for-10th-lowest-on-record.

Neimark, Peninah, and Peter Rhoades Mott, eds. *The Environmental Debate: A Documentary History.* Westport, CT: Greenwood Press, 1999.

Nelson, Gaylord. *Beyond Earth Day: Fulfilling the Promise.* Madison: University of Wisconsin Press, 2002.

Ortiz, Joseph, and Roland Jackson. "Understanding Eunice Foote's 1856 Experiments: Heat Absorption by Atmospheric Gases." *Notes and Records: Royal Society Journal of the History of Science* 76, no. 1 (August 26, 2020): 67–84. doi.org/10.1098/rsnr.2020.0031.

Plumer, Brad. "Climate Change Is Speeding Toward Catastrophe. The Next Decade Is Crucial, U.N. Panel Says." *New York Times,* March 20, 2023. nytimes.com/2023/03/20/climate/global-warming-ipcc-earth.html.

Renaud, Jean-Paul. "Concrete Mountain to Be Razed." *Los Angeles Times,* June 16, 2004. latimes.com/archives/la-xpm-2004-jun-16-me-mountain16-story.html.

Roosevelt, Mark. "Theodore Roosevelt's Great-Grandson Says: Remove the Statue." CBS News. July 12, 2020. cbsnews.com/news/theodore-roosevelts-great-grandson-mark-roosevelt-says-remove-the-statue.

Schwartz, John. "Overlooked No More: Eunice Foote, Climate Scientist Lost to History." *New York Times,* April 21, 2020. nytimes.com/2020/04/21/obituaries/eunice-foote-overlooked.html.

Shapiro, Maura. "Eunice Newton Foote's Nearly Forgotten Discovery." *Physics Today.* August 23, 2021. doi.org/10.1063/PT.6.4.20210823a.

Shiva, Vandana. *Staying Alive—Women, Ecology, and Development.* London: Zed Books, 1988.

Thoreau, Henry David. *The Maine Woods.* Boston: Houghton Mifflin, 1893.

Tyndall, John, 1859. "Note on the Transmission of Heat Through Gaseous Bodies." *Proceedings of the Royal Society of London* 10 (January 1, 1960): 37–39. doi.org/10.1098/rspl.1859.0017.

Uekoetter, Frank, ed. *The Turning Points of Environmental History.* Pittsburgh: University of Pittsburgh Press, 2010.

Zhong, Raymond. "Arctic Summer Could Be Practically Sea-Ice -Free by the 2030s." *New York Times,* June 6, 2023. nytimes .com/2023/06/06/climate/arctic-sea-ice-melting.html.

IMAGE CREDITS

Joseph-Siffred Duplessis, National Portrait Gallery/Wikimedia Commons (p. 13); Lisa Hupp/U.S. Fish & Wildlife Service (p. 16); Lewis Hine, National Archives at College Park/Wikimedia Commons (p. 20); Elmer Ellsworth Burns, 1910. *The Story of Great Inventions,* Harper & Brothers, New York, p. 39, fig. 13 (p. 21); Josiah Johnson Hawes, 1857/ Wikimedia Commons (p. 24); "How We Get Gold in California" by A Miner of the Year '49. "River Operations at Murderer's Bar," *Harper's New Monthly Magazine,* April 1860, p. 603 (p. 26); Federal Highway Administration, 1975/Wikimedia Commons (p. 27); Thomas Cole, "View from Mount Holyoke, Northampton, Massachusetts, after a Thunderstorm—The Oxbow," 1836/Wikimedia Commons (p. 30); "Hon. George Perkins March of Vermont," circ. 1855–1865, Brady-Handy Photograph Collection, Library of Congress, LC-BH82-4981 A (p. 45); Thomas A. Ayres, "The First Picture of Yosemite Valley," June 27, 1855/Wikimedia Commons (p. 46); Carleton E. Watkins, 1883/National Portrait Gallery, Smithsonian Institution (p. 48 top); Carleton E. Watkins, "View from Inspiration Point," 1879, Princeton

University Art Museum/Wikimedia Commons (p. 48 bottom); Carleton E. Watkins, "Yosemite Falls, 2,634 feet," circ. 1872, printed circ. 1876, Metropolitan Museum of Art/Wikimedia Commons (p. 49 top); Carleton E. Watkins, "Section of the Grizzly Giant, Mariposa Grove, Yosemite," 1861, California Historical Society/Wikimedia Commons (p. 49 middle); Carleton E. Watkins, "Piwyac, Vernal Fall, 300 feet, Yosemite," 1861, National Gallery of Art/Wikimedia Commons (p. 49 bottom); Thomas Moran, "The Grand Canyon of the Yellowstone," 1872, Smithsonian American Art Museum lent by the Department of the Interior/Wikimedia Commons (p. 50); © Kate Messner, courtesy of the author's personal collection (p. 51); "'Forest and Stream's' Yellowstone Park Game Exploration: The Account of Howell's Capture," *Forest and Stream,* May 5, 1864, p. 377 (p. 56); National Park Service, 1894 (p. 58); © 1904 by Pach Bros., "Theodore Roosevelt, Pres. U.S. 1858–1919" Library of Congress, LC-USZ62-88858 (p. 60); LunchBoxLarry, "Equestrian statue of Theodore Roosevelt"/Wikimedia Commons (p. 62); "The wire mill, Donora, PA" © by Bruce Dresbach, Donora, PA, 1910, Library of Congress, LC-USZ62-131258 (p. 71); "First Marines Mortar Crew, Guadalcanal, circa 1942," Archives Branch, USMC History Division/Wikimedia Commons (p. 72); "Shoot to Kill—Protect Your Victory Garden" 1941–1945, Office for Emergency Management, Office of Information, Domestic Operations Branch, Bureau of Special Services/National Archives (p. 74); "DDT is good for me-e-e!," *Time Magazine,* June 30, 1947/Courtesy of Science History Institute (p. 75); NCTC Archived Museum, Rachel Carson Collection, 1952/U.S. Fish and Wildlife Service (p. 81); Marion S. Trikosko, April 20, 1979/Wikimedia Commons (p. 84 left); Jay Godwin, April 8, 2019/Wikimedia Commons (p. 84 right); John Malmin, *Los Angeles Times,* July 9, 1975/Wikimedia Commons (p. 84 bottom); "Cuyahoga River fire 1952 taken from the Jefferson Avenue Bridge," November 1, 1952, The Plain Dealer © 1952 The Plain Dealer. All rights reserved. REPRINTED/USED with permission. (p. 90); White House

Photo Office, U.S. National Archives, December 31, 1970/Wikimedia Commons (p. 100); United States National Museum of Natural History, Natural Collections, Cincinnati Zoo/Smithsonian Libraries and Archives (p. 106); Ron Garrison, San Diego Zoo/Wikimedia Commons (p. 107); "Love Canal Pre 1982 Ring 1 Homes," Environmental Protection Agency (p. 115); Jimmy Carter National Historic Park, Georgia, National Parks Service, December 11, 1980 (p. 117); Hazel M. Johnson, courtesy of the People for Community Recovery (p. 129); Ben Cody, September 17, 2011/Wikimedia Commons (p. 132); U.S. Coast Guard (p. 133); Exxon *Valdez* Oil Spill Trustee Council (p. 134); Meisenbach Riffarth & Co. Leipzig., 1909/Wikimedia Commons (p. 152); National Science Foundation, June 12, 2022/Wikimedia Commons (p. 154 top); Ivtorov, "The Mauna Loa Observatory"/Wikimedia Commons (p. 154 bottom); Dr. Pieter Tans, NOAA/ESRL and Dr. Ralph Keeling, Scripps Institution of Oceanography. Accessed 2023-12-15./Wikimedia Commons (p. 155); "Photograph, Warren Washington and Akira Kasahara," 1975, Warren Washington Papers, NSF NCAR Archives (p. 156); U.S. Coast Guard, April 20, 2010/Wikimedia Commons (p. 164 top); NOAA's National Ocean Service, June 14, 2010/Wikimedia Commons (p. 164 bottom); Fibonacci Blue, October 25, 2016/Wikimedia Commons (p. 167); NOAA Fisheries, November 8, 2018 (p. 170); Scott Schliebe, U.S Fish and Wildlife Service/Wikimedia Commons (p. 171 top); USDA/Wikimedia Commons (p. 171 bottom); NASA, MODIS/LANCE, HDF File Data processed by Supportstorm, October 25, 2012/Wikimedia Commons (p. 172 top); Pacific Southwest Forest Service, USDA, September 25, 2020/Wikimedia Commons (p. 172 bottom); Photo courtesy of Dr. Colin Donihue, Washington University, St. Louis, MO (p. 196)

INDEX

C

D

E

F

G